# Harvest for Profit

## An Essential Guide to Growing Low Maintenance Fruit

By Ruth Linton

ISBN 978-1-959987-02-4

www.highlandorchardsfarmmarket.com

# Table of Contents

# Preface

Why do you want fruit that you can ignore until it's time to harvest? Three main reasons.

First, the farmer's to-do list is always longer than the time available.

Second, no one wants to deal with sprays or chemicals.

Third, extra income without a lot of extra work is always welcome.

This is the first in a series of concise books for small farm profitability. It is possible for small farms to bring in big profits, with the help of crops that are not labor-intensive. My family has farmed fewer than 100 acres for nearly 200 years. For the past 70 years, we have maintained the core 5.5 acres. We grow fruits, vegetables, and flowers, and we sell directly to the public. The farm currently supports three generations plus several employees.

This knowledge of how to approach growing as a profitable enterprise is what makes this book different. We are looking at ways to make small farms more profitable without bringing on a lot more work.

Now if you are growing for the fun of it and you have lots of time to fuss over plants, then you may have other fruit on your list of things to grow. However, you just may want to peek at these plants because they have fabulous flavor and are a great addition to any home garden as well as for the professional farmer.

If you want to add fruit to your farm or homestead or back yard, and you don't want to spray or spend massive amounts of time in maintenance…

And if you want to grow the income of the farm without adding hundreds of hours of work…

Then fruit that requires very little maintenance is a good fit for you.

We have always grown fruit here on our family farm. In addition to loving fruit for its flavor and nutrition, we have found that it is a good income stream for the farm. Certain fruits command a premium price and will attract customers to you just for those fruits.

We have worked with tree fruit, berries, and shrubs. We know the ones that are hard to grow, hard to harvest, hard to store, and hard to sell. We also know the ones that are easy to grow, easy to harvest, easy to store, and easy to sell.

Here I take you to our top five low-maintenance, easy to grow organically, stores well fruits. I know! Look at that list—

- Low-maintenance fruit?
- No spray fruit?
- Fruit that stores well?

Yes! In fact, many people will say that these fruits do not store well. However, there are tricks to harvesting at the correct time which make it possible to extend the shelf life of all of these fruits.

I am not saying these are the only low-maintenance fruits that you can grow. Depending on your particular area, there can easily be other fruits that work well for you. I want you to look at these fruits with a different set of criteria.

Often, we look at a new crop and think,

"That sounds good."

"I bet I can grow that."

"That is delicious, we should grow that."

This is more a model of how the fruit can help support your farm as a business. You want the crop to reward you for your time invested in it. The criteria then become:

How much work do these plants require?
Do these plants fit in with my farm and what I already do?
Will I be able to store the fruit until I can sell it?
Will I be able to sell this fruit at all?

Since these are perennial plants, then you are making an investment by planting these in your valuable ground. If you are growing crops to pay the bills and to make some money, then this is information you want to have.

# Introduction

My family has grown fruits and vegetables on a small farm since 1832.

This is the main road leading to the farm. In 1900, Foulk was a dirt road, and now it is a four-lane paved highway. Change and more change! However, we are *still* growing fruits and vegetables, herbs, and flowers.

Twenty-five years ago, I started looking for items to add to our mix. Specifically, items that had a longer shelf life than the raspberries we were growing, and that were not as fickle about the weather. Crops that did not need spraying and crops that were appealing to our customers. Currants, gooseberries, figs, persimmons, and pawpaws are the result. We were already growing two of them but in very small quantities. We have been growing these five crops ever since, and they are a great addition to our selections, marketing, and profitability. No one else grows these in our area and no one else sells them farm to table.

Growing is what we do and how we pay our bills.

There are lots of people growing a wide variety of things and talking about them. What I offer you is the farmer's point of view as to how these crops can fit in with the business side of farming. I focus on how these plants can help improve your profit.

# Myth Busting

## The negative voices around you and, maybe, in your head

*"You cannot grow fruit organically."*

To achieve best appearance, many types of fruit do require some sprays. These are the fruits that CAN be grown without *any* sprays! They have a great appearance for retail sales without any fuss.

*"Growing fruit is too complicated."*

It is true that many fruits do have a strict regimen of pruning, thinning, fertilizers, and sprays, including organic treatments. However, not these fruits! These fruits are hardy, and just need a dose of compost once or twice a year.

*"Fruit trees take a lot of time."*

If you are thinking of peaches or apples, indeed, there are multiple steps involved. Many fruits do need pruning, spraying, netting, fertilizing, thinning, trellising, or other attention. That is not true for this group! The ones we are discussing are the easy care, once-a-year attention, and then ignore till harvest. Sound good?

*"What if it doesn't grow in my growing zone?"*

I will give you tips and tricks for growing beyond the official growing zone. The first tip is there are varieties for extending the zones and getting what you want. And if it isn't for your growing zone, I will tell you why.

*"I don't know anything about growing fruit. I've only ever grown vegetables."*

I agree that there are many fruit shrubs and trees that need a special skill set. Some fruits are high maintenance and have a very short shelf life. That's the reason I want you to know about these specific fruits. You already possess the basic skills of growing. It is not a stretch to incorporate these five into your knowledge base. We will learn about some farm-favorite fruits, which will grow easily, without fuss, AND then sell profitably for you.

Remember, the goal in all of this is for you to find a way to bring more profits to your farm with fewer work hours, not more work. If a crop is not a good fit for you, then you now know to avoid it.

# Why listen to a farmer about growing?

## Good question!

There is a lot of advice available in the world through books, videos, websites, and social media. There is information all over the place. Most of this information is valuable. But there is an additional way to look at this information.

Farmers have a different point of view. Farmers look for maximum results with the least amount of time involved as possible. There are always plenty of projects to fill up the time. What we need are crops that take care of themselves until it is time to harvest.

There are so many demands on our time, that the list seems to only get longer rather than finished. Having crops that we can ignore until harvest is a gift to the farmer. Time is a precious commodity, and we usually don't have enough of it to do everything that we want.

Because we make a living from selling what we grow, we look for high quality and excellent quantity as a return on the investment of our time, land, and money. Every square foot that is planted needs to help pay the bills. As a result, we want to create the growing areas with the most abundant and profitable crops.

Farms have their own type of beauty because they are filled with lots of plants and flowers. But the goal is to make those plants and flowers work for us, rather than for the people to work for the plants.

As much as we love growing and love plants, we want those plants to help support us financially. Sadly, simply loving plants does not pay the bills.

In addition to all the other great attributes of these plants, these five low-maintenance fruits provide reliable crops. You can depend on having fruit year after year. We have had one year in 50 years when we had no figs. Figs for 49/50 years. Currants for 50/50 years. Gooseberries, persimmons, and pawpaws every year since they started fruiting, 20 years. All of that with very little attention!

# Five fruits and what they have in common

Here are the basics in common for all of these plants:

**A place to grow**

These are all perennial shrubs and trees, so you want a spot where the plants can sit undisturbed. These all live for many years and some can live for well over a hundred years. Fortunately, none of them were very fussy about location. Most want all the sun you can find for them. Currants, gooseberries, and figs can handle partial shade and still do well. If you get hot weather, over $85^F$ in May or June, then your gooseberries or currants need to have some afternoon shade. Figs will still produce well with some shade. However, just as with most crops, sun up to sun down will give you the most figs per tree.

Planting Location: Sun, adequate water, good drainage, access to nutrients

First choice is full sun, the farmer's definition of full sun: dawn to dusk. Gooseberries, currants, figs, and pawpaws can handle only 8 hours of full sun and partial shade in the afternoon, although you have a more abundant crop with full sun. The figs and pawpaws will grow a bit taller as they reach for more sunshine with only 8 hours of sun. Persimmons need full sun in order to produce abundant fruit.

Be sure to mark the location with some stakes, so that the young plants (they just look like sticks in the ground for quite a while) do not get weed-whacked or mown down. All of these thrive in average soil which has proper drainage.

Mulch is great to add around the plant. It helps suppress weeds and to hold moisture in for the young plant. Add about two inches of mulch. We have found the mulch has three benefits: it suppresses weeds, holds in moisture, and breaks down into compost for nutrients for the following year.

We use dried leaves as mulch. We collect mounds of leaves in the fall for use each spring. If you do not have lots of your own leaves, you can offer a landscaper a place to dump leaves on your property. This saves the landscaper a trip to the dump and the associated fees. You can either charge a small fee, or, allow them to dump for no fee and build up credit for when you might need a favor. We use leaf mulch all over our farm.

When you mulch, make a ring of mulch around the plant, about one foot in all directions, and about two inches deep. This helps suppress weeds and to retain moisture in the soil for the young plant.

You want the terrain to be relatively flat. I say relatively because we do not have a level section on the farm. Everything slopes a bit. That helps with drainage. But a steep hill will make it difficult for any of these plants to get sufficient moisture. Look for the places that have only a few degrees of slope, about one-foot drop per hundred feet, no more than two feet.

We do not irrigate any of these plants unless it is a drought year. All of these plants do well if they get about ½ inch to an inch of water per week on average. If your rain does not average out to that much, then consider setting out drip irrigation and a timer.

**Plants**

You will most likely get your plants from a non-local nursery. If you have a local nursery that stocks these, I am jealous. I would love to be able to check out the shrubs and trees in person.

Nurseries have worked out how to ship very well, so you do not need to worry about shipping. The only tricky part is dealing with the plants immediately once they arrive. You cannot delay unpacking and planting your trees and shrubs.

Most of the shrubs and trees will come as bare root stock, meaning they look like sticks with some roots. Generally, the roots are wrapped to hold moisture during shipping. When you receive your plants, you want to get them in soil as quickly as possible. If you cannot plant right away, then plant them in large pots and water well. Transplant into the ground within two weeks.

The exception to receiving bare root stock will be a pawpaw tree. Pawpaw trees have a long taproot with lots of small fine roots attached, and they dislike being disturbed. Your pawpaw tree should come in a tall, skinny pot which protects the taproot. If you deal with a small nursery for your pawpaw trees, they may not ship and will require pick up at a certain time.

Pawpaws should arrive in pots. Do not transplant into a larger pot. Pawpaws should go in their final location as soon as possible, as they do not like to be transplanted.

**Nutrients**

Aged manure and compost are the best types of fertilizer for all of these plants. No special pH requirements are needed, just moderate from 5-7. If you do not already have a compost pile, I encourage you to start one. Compost is well worth the time.

If you are using manure, make sure it is aged at least six months. Fresh manure can burn the roots of plants. Because aged manure is potent, it is best to mix the manure with some soil and then add it as a top dressing, and not directly into the hole as you plant the bushes or trees. The manure can burn the roots and stunt the growth of your plant. Bedding mix or potting mix will be a mixture of compost, top soil, and peat moss, and can be added to the planting hole as is. Mix with the soil you dug out and place around the roots.

Compost should be about one-third of the soil around the plant's roots. Manure is a potent fertilizer, so only a small amount is needed. A single shovelful spread around the soil on top at planting is all that is needed. Manure should not be put on the roots; simply top dress with a shovelful around the plant. After planting, do a top dress of compost each year in the spring. Too much fertilizer at planting will spike top growth before the roots have had a chance to settle in and grow. You want the roots large enough to support the top growth, so the roots need to grow first.

In general, be careful that any manure you use has been aged at least six months before you use it around the plant. It is great fertilizer and well worth the wait.

Each spring, add some compost and leaf mulch around each plant. One or two shovelfuls of compost per plant is enough. That is sufficient food for the season. Top it with some leaf mulch to suppress the weeds.

The exception is your fig tree. If you can add compost or aged manure in the middle of the season, usually in July, the fig tree will thank you with additional figs. If you forget, the fig tree will still give you lots of figs.

**Water**

Water when planting. In the first month or two, the plants are growing roots, and you do not see a lot of top growth. Make sure the roots have water! If you remember to water the entire first year, then you have done exceptionally well. I confess, we water for the first month or two, and then the rest of the farm demands our attention. The plants all seem to forgive us.

As with any first-year plants, a little extra care during the first growing season helps set the plants up well. Maintain your water schedule, so that the new plants get water each week.

You will need to water diligently in the months after planting until the plants' natural dormant season. In the years following, unless you have a prolonged dry and hot period after that—no rain for a month, you will not need to water. If it rains that week, no need to water additionally.

For young plants, a two-inch layer of mulch can help keep the soil from drying out as quickly.

We have found that mulch works wonders for moisture retention during dry, hot times. It is well worth the time in late winter/early spring to put fresh mulch down.

Just as with all the other requirements, the water needs are moderate for each of these. Although we have drip irrigation for all the vegetable crops we grow, we have found that we normally do not need to irrigate any of these crops past the first year. You can set drip irrigation down for the first year to make sure the plants have water regularly. If you have just a few plants and you do not have irrigation at this point, water at the root level each week. Make sure the water reaches the bottom of the roots.

Drip irrigation is by far the most efficient way to water during any dry period. If you only have sprinklers or you are watering by hand, be sure that you water long enough for the deepest roots to receive water. Avoid water getting on the fruit when it is close to maturity.

If it is an extra dry year, pushing towards drought, then we will water the shrubs—gooseberries, currants, and figs. The trees have long enough roots to tap into moisture deep in the soil.

Planting where the shrubs or trees will have good drainage is important. None of these want to be where there are puddles or standing water. Rain can help increase the size of the fruit, but as long as excess rain can drain off the roots, you will not have a problem.

**Storage**

All of these fruits require refrigeration for storage after harvesting. Left at room temperature, you will have 1-2 days in which to use or sell the fruit. In proper refrigeration, 32-36 F., you will have 7-30 days in which to sell or use the fruit, depending on the crop.

Cardboard containers are best. You do not want closed plastic containers.

**Find your plant**

These are the nurseries where we order. There are hundreds, if not thousands, of nurseries that offer all kinds of plants. Wherever you find your plants, make sure the nursery offers disease-free stock.

Pense Berry Farms www.penseberryfarm.com Gooseberries and currants

Nourse Farms https://www.noursefarms.com Gooseberries and currants

Willis Orchards https://www.willisorchards.com Pawpaws and persimmons

Ison's Nursery https://www.isons.com Persimmons

Check with your state Department of Agriculture or extension agent regarding the legal status of currants and gooseberries in your area. Gooseberries and currants can be the host for white pine blister rust, which does not kill the *Ribes* but which can kill white pine trees. If there is a significant pine tree industry in your state, you probably will not be able to grow currants or gooseberries. Regardless, it's important that you get disease-resistant varieties and healthy stock to begin with. Sometimes, some varieties will be prohibited but the other varieties are allowed.

Look for varieties that are suitable for your growing zone.

**Currants**:

Zones 4-8 (we are zone 7 and grow them). The variety matters a lot, here! Zones 8-9 will probably be too hot for most varieties. However, currants are very cold hardy.

**Gooseberries**:

Zones 4-7 (we are zone 7 and grow them easily). Zones 8-9 may be too hot. Zone 8 may work with some specific varieties and afternoon shade. Gooseberries tolerate cold to -40 F.

**Figs**:

Zones 5-9 (Yes, I know farmers who grow figs in Zone 5!). *Chicago Hardy* is a variety more suited to colder winters, hardy to -20 F. Winter protection is key.

**Pawpaws**:

Zones 4-9, several different varieties.

**Persimmons**:

Zones 4-9, depending on the variety. The variety is key to finding the right tree for your zone.

# Currants and gooseberries

*Ribes* is the genus in the family "Grossulariaceae" which encompasses currants and gooseberries. There is a lot of argument about the classifications of the 200 or so shrubs within this, but we are not going to worry about the arguments. We are sticking with growing these two types of fruit.

Gooseberries and currants are both deciduous shrubs, dropping their leaves each fall, and both are cold-weather hardy to -40$^{F}$. Both survive quite well on their own in uncultivated areas. Wild gooseberries and currants can grow in wooded areas, but there will be little fruit available. Sunshine is necessary to get fruit in abundance.

Gooseberries have been cultivated in northern European areas for so long that it's difficult to distinguish between the original wild strains and the cultivated strains. In Scandinavian countries, gooseberries are grown as far north as the outer edge of the Arctic Circle. They have been treated as a hedgerow (using those thorns for natural barriers), as food for wild animals, and cultivated by humans for fresh eating, jams, pies, syrups, and wine.

Currants have been grown throughout England, Scandinavia, Europe, and western Asia for centuries. Red and white currants have been highly prized in France for Lorraine jelly, and red currants are the most commonly used filling in Austria for Linzer tortes. Currants are also popular as a dried fruit, and then added to scones, buns, and fruit bread. The leaves have been used to infuse with black tea as well. Currant syrups and wines are popular in western Asia.

Red currant jelly is a popular condiment in Britain, served with lamb, venison, turkey, or goose.

Currants and gooseberries arrived in North America along with immigrants in the $17^{th}$ century. The *Ribes* adapted easily to the eastern half of the continent.

Gooseberries used to be plentiful throughout North America and were as popular as blueberries throughout the $19^{th}$ century. Then, about 100 years ago, scientists discovered that they are a host to a mildew called the white pine blister rust. This rust needs two hosts to develop, *Ribes* and white pine. The timber industry succeeded in having federal and state governments ban gooseberries and currants from areas near the tree stands. Gooseberries nearly disappeared from the North American landscape for most of the twentieth century.

New varieties that are mildew resistant have made it possible to grow gooseberries again in most states in the United States. Other countries do not appear to have these same restrictions, as they do not have a pine tree industry that is affected.

My family used to grow rows and rows of currants in the 1910s and 1920s. They took crates of red currants (*Ribes rubrum*) to the farmers' market in Wilmington, selling to all the people who then made jelly, syrups, and homemade wine. Excess currants were shipped to Philadelphia for the markets there.

We started growing gooseberries after I had customers ask about them. I looked into how to grow them, and they seemed like a good fit for our farm. We have expanded from three plants to twenty to forty plants.

The federal government ceased its ban on growing gooseberries and currants in 1966. Individual states may still have restrictions, particularly in the northeast.

Some states have realistically acknowledged that white pine blister rust already exists in the state. Since the newer disease-resistant varieties of currants and gooseberries are the ones now cultivated, the *Ribes* will not have a negative effect on white pines in the area.

# Varieties

## *Currants varieties*

Red currants, *Ribes rubrum,* are the most common type of currants grown. They are hardy, reliable, and prolific in growing. The bushes grow to three to four feet in height, and about two to three feet wide. The fruit grows in clusters looking like miniature clusters of grapes. If there is a ban on currants, usually *Ribes rubrum* will still be allowed to grow. Very few areas prohibit this variety.

There are actually red, pink, and white berries that are all *Ribes rubrum,* depending on the specific variety. Red currants can be traced to France in the 12th and 13th centuries.

**Fertile de Palluau** is an excellent variety with moderately long bunches, dark red color, and a sweet/tart combination of flavor. It matures in late June, and works well for growing zones that develop hot weather in July and August.

**Mulka** is a red currant that matures in late July with fruity flavors and tight clusters. This variety is better suited to growing zones with milder summer temperatures.

**Junifer** is a good variety that matures in early June. Its early harvest time is good for growers in zones with milder winters and hotter summers. Excellent flavor with a good balance of sweet and tart.

**Red Lake** is a classic red currant, with large, fairly loose clusters, and a little more sweet than tart in flavor. Harvest in late June or early July.

**Pink Champagne** is an American variety similar to *Gloire des Sablons* but with added disease resistance to mildew. Harvest in early July.

**Gloire des Sablons** is a beautiful pink currant of French origin with a sweet and acidic flavor. It is prized for making a beautiful pink jelly. It matures in early July.

**Black currants**, *Ribes nigrum*, are the currants most likely to be banned if your state or county has restrictions. However, the breeding programs in many countries have developed fruit with great hardiness and disease resistance. If you can incorporate Black currants into your growing program, it is well worth the investment. Black Crandall is the best of the black currant varieties.

Black currants will grow to about five feet by five feet, making them larger than their red and pink siblings. The plant is very aromatic, and the berries have a stronger flavor than the red currants. In general, black currants are harvested a few weeks later than red or pink currants.

Black currants are commonly used in cooking rather than in eating fresh, as the strong flavor does not appeal to most palates. Black currant cordials or black currant "squash," both of which are a type of fruit drink, are the most popular use for black currants in Europe. Black currant jam and sauces are frequently paired with meats, particularly wild game.

Black currants were brought from England with the settlers in 1629 in the Massachusetts Bay Colony, and have been here ever since. In 1919, just before the federal ban, New York produced 3.3 million quarts each year, about half of the annual production in the United States. Black currants were the last variety of *Ribes* still forbidden in New York State, and the ban was finally lifted in 2003.

Each variety of currants is harvested within a 7-10 day period. You can extend your harvest season by growing different varieties.

## *Gooseberry varieties*

The European variety of gooseberry is *Ribes uva-crispa,* and the fruit are usually large, sometimes as large as a plum. These varieties naturalize well throughout Europe, but are susceptible to mildew and rust. Oftentimes they are referred to by the botanical name of *Grossularia* instead of "gooseberry." European gooseberries are not grown in the United States due to the mildew issues and that they are a host to white pine blister rust.

The North American species are *Ribes hirtellum,* which have smaller berries than their European cousins and a smooth skin. There is also an ornamental flowering gooseberry, *Ribes speciosum,* with gorgeous fuschia flowers and no fruit.

The following are varieties suited for North American growing:

**Pixwell** is a very hardy variety with large, dark pink/purple berries. This is a productive variety that sets abundant fruit. The thorns are much less pronounced in the bushes, and the berries have an excellent semi-tart flavor for pies and jams.

**Black Velvet** is a hybrid of gooseberry and Worcesterberry, and has excellent commercial production potential. The dark red fruit is sweeter, closer to a blueberry in taste, with high quality and heavy cropping. It is self-fertile, and mildew and disease-resistant.

**Captivator** is a US-European hybrid that has good resistance to mildew, excellent flavor picked directly off the bush, and an attractive blush pink color.

**Champion of Oregon** is a west coast hybrid with greenish-yellowish berries. Suited for growing in the northwest, the plants are medium to large in size. They have a tart flavor which is good for commercial use.

European varieties that grow well in North America:

**Hinnonmaki** is a Finnish heirloom variety with medium-sized berries. The skin is tart and the interior flesh is sweet, which provides a captivating flavor sensation. The berries can be red or yellow. The red seems to be the favored flavor. It has a long harvest season in its native area and is disease resistant. This could have a long harvest season in zones 4-5 in North America as well.

**Invicta** is resistant to mildew and is very hardy. It yields a large crop, but the berries have a milder flavor. Gooseberry aficionados do not rate the flavor highly, but the good disease resistance, hardiness, and heavy cropping make it a popular variety. We have found it sells well with retail customers.

**Leveller** is a large-fruited plant, with berries as large as strawberries and blackberries. The berries are yellow and have excellent flavor, prized for eating fresh and for cooking. It has greater disease resistance than most European varieties and is the preferred variety in Great Britain.

There are many varieties of gooseberries available, so check the details carefully. Disease resistance to mildew is critical. Once the plant has mildew, there is no cure.

# General planting and care

Gooseberries and Currants are relatively easy to grow. Their main issue is they don't like the temperatures too hot. Once you know this, it makes sense that these plants do well in Scandinavia, northern Europe, England, Canada, and the northern half of the United States.

For those of us with hot summers, we have to exert a little care in selecting our varieties and planting location. If you have more than 5 days above 90 F in the summer, then afternoon shade is a must. Selecting varieties that mature earlier is also helpful, as the fruit ripens before the extreme heat of the summer hits.

Our currants and gooseberries ripen in late June, usually just before we hit our very hot weather. We are growing zone 7a. We have them planted where there is some afternoon shade. For a farm, partial shade usually renders the space less useful, as nearly all crops prefer sun up to sundown sunlight. However, for the gooseberries and currants, they set the fruit in the spring before the surrounding trees leaf out, thus getting the benefit of the full sun in the cooler months. As the tree leaves grow, the bushes get some shade, which reduces the temperature a little bit.

This way, we are able to plant some crops that produce income in spaces that are not conducive to most vegetables or fruits.

If you are in zones 6 and cooler, then full sunshine will give you larger fruit and a bigger harvest. Gooseberries tolerate a wide range of soil types and moisture conditions. They will perform exceptionally in nutrient-rich soil with good drainage. However, we have found they grow well even in our heavy clay soil. In sandy soil, adding compost will be essential to provide enough nutrients for the plants to develop fruit

They need 40 days of winter chill below 45 F. Zones 4-8 (we are zone 7a and grow them) can support these plants. Check the specific variety for hardiness to cold and to heat. Most gooseberries and currants can tolerate winter temperatures to -40 F.

# How to grow

If your *Ribes* are shipped from a nursery, they look like sticks with some roots. This is called "bare root stock." The roots are generally wrapped in damp shredded paper. This is completely normal. The currants and gooseberries are in a dormant state, and the roots are kept damp to keep the plant alive.

You want to plant your currants and gooseberries in the ground or in a pot right away. It is important for them to establish root growth before the summer heat begins. If you cannot plant right away, keep the young plants refrigerated. You can keep them refrigerated for up to two weeks. After that, you risk developing mold on the roots.

If you purchase your currants or gooseberries at a local nursery, then most likely they are already in a pot.

The sooner you can get your plants in the ground, the sooner they start establishing their root system and will be ready to come out of dormancy, usually very quickly.

For bare roots stock, dig a hole at least twice the depth of the roots and about two feet in diameter. For anything in a pot, you want the hole to be twice the diameter of the pot and twice as deep. You want plenty of space for the roots to spread out. Fill the hole with a good mix of compost and the soil from the hole. If the soil does not have many nutrients, replace it with a bedding mix or compost mixed with the existing soil. Compost can be about one-third of the soil mix around the roots. If you are planting a row of plants, we find it easy to dig a trench and then fill in around the plants.

After you plant, water deeply so that the water reaches as far as the tips of the roots. This can take several minutes for the water to soak in. Plan to water weekly for the first summer. If it rains about a half-inch, then you do not have to water that week.

Space your plants about four feet apart in a row, and the rows should be eight feet apart. Good air circulation is essential for plant health. The plants will fill in.

The first year, you will not have any fruit. Gooseberries and currants start producing on canes that are two years old. These plants produce fruit on canes for three years, and then the canes die. Every year, you will have new canes, one-year-old, two-year-old, and three-year-old canes. In the spring, you will see canes that have new leaves sprouting, and several that are completely dead.

It's an easy clue for which ones to prune out (the dead ones).

In pruning, the goal is to have 3-4 canes of each year on the plant. Older canes then die out and need to be pruned off. Gooseberries bear on one-, two-, and three-year-old canes. Annual pruning keeps the plants healthy and helps with air flow.

Fertilizer—gooseberries and currants do well with some aged manure or compost spread around each plant in the spring or manure mixed with compost. This is the nutrient-rich part that these plants prefer. Since our farm has a clay soil base, we add compost to all plants every year.

Air circulation is important, so plant 4-5 feet apart to allow for growth and air flow.

Gooseberries and currants need 40 days of winter chill, which is under 45 F. All of the *Ribes* tolerate winter temperatures to -40 F. We have not had a problem with winter die off. These plants define winter hardy! If your temperatures are below -40 F or you have severe winds, you may want to add a wind break or cover with row cover for a little protection. Remove the cover before the plants come out of dormancy.

Mulching is a good idea, as these *Ribes* have fairly shallow roots. They do not like competition from weeds and they like their roots protected.

These plants do not do well in extreme summer heat. We "officially" are a zone too hot for several varieties of currants and gooseberries. We have found that they prefer afternoon shade to protect the plants from the summer heat. The current location provides shade from about 2:00 on, and the plants are doing well. In the previous location, the plants had full sun, and they struggled. Because of the shade, we get about 2/3 of the harvest than if the plants were in full sun in a cooler zone.

When the currants were in full sun here, the berries would start to dry on the bush, and the leaves would yellow early in the summer. The plants did not look healthy. Once transplanted to a cooler spot on the farm, with a little shade, the plants revived and did much better. The gooseberries were fine in full sun, and have done as well in the partial shade area as well.

On average, we get about 10-12 days of temperatures over 90 F. Last year we had 30 days over 90 F. Because the plants have afternoon shade, they came through just fine. This year's crop looks good.

Like everything else, currants and gooseberries like well-draining soil. Mulch, about two inches deep, helps protect the roots from heat and helps retain moisture. We rarely water the plants.

Although the mature berries are heavy, we have not found it necessary to stake the plants. If you set up some stakes, you can drape bird netting over them to protect the ripe fruit from birds. In general, we are able to harvest before the birds so we don't use netting.

The bushes grow to about 3-4 feet tall, an easy height for picking. If you want a windbreak or a hedgerow at the edge of your property, gooseberries make for an easy hedge. Plant them closer together, about 2 feet apart, to form a thick inter-twining row. Their thorns help deter unwanted visitors, both two-legged and four-legged. Currants can also provide a windbreak, although they do not have the thorns of gooseberries. Some of the newer varieties of gooseberries are thornless, so if you want a thorny hedge, choose your varieties carefully.

Currants and gooseberries will start to produce fruit after one year, and will produce full crops after 2-3 years. The plants live for decades, as they continually put out new fruit-producing canes.

# Maintenance

Currants and gooseberries are a great addition to the home garden or small farm to eat fresh or to make pies, jelly, sorbet, or jam. They grow easily, not needing too much special care, and rarely need staking or supports. They do not need sprays, organic or otherwise. Gooseberries, in particular, can make a great natural hedgerow or windbreak or deterrent (because of their thorns), plus you get fruit.

If you have high levels of birds in your area, you may need to net them. You can build a simple frame and tack the net onto the frame, over the bushes. We do not net. We find that if we harvest at the right time, we do not have an issue with the birds. If you allow ripe fruit to stay on the bushes, then the birds take that as an invitation.

Check to make sure the varieties you get are resistant to powdery mildew. The mildew is what leads to the white pine blister rust. There is no cure for mildew. The mildew will eventually kill the *Ribes*.

American varieties of gooseberries are usually hybrids of the European varieties. The European varieties are more susceptible to mildew. For gooseberries, we like Black Velvet (if allowed in your area), Invicta (zones 3-7), Captivator, and Pixwell, and have had zero problems with mildew. For currants, we grow Red Lake Currant (*Ribes sativum*), and Pink Champagne Currants (*Ribes rubrum*), also hybrids. Black Crandall is a great black currant if allowed in your area.

# Harvest

The ease of growing is offset by the challenge of harvest and cleaning for use. Gooseberries have significant, sharp thorns, up to an inch long on the older canes, and are quite hard. The thorns can do serious damage. You always want to pick berries with long sleeves and long pants.

Although currants do not have thorns like their cousins, gooseberries, they have fine hairs on the leaves and branches which can irritate your skin.

Picking the berries is much like harvesting blueberries or blackberries. The stems are very tender and the berries slip right off. Harvest into soft cardboard boxes. The standard berry box is ideal.

Our harvests occur from about June 20 through July 20. You harvest each variety over the course of about ten days, so different varieties will give you an extended harvest. Cooler zones will have a longer harvest period as the bushes will keep producing in their favorite weather. Gooseberries tend to ripen just after the currants.

Get the berries into refrigerated storage as quickly as possible. This extends the shelf life of the fruit. The berries will continue to ripen slowly in refrigeration.

For cooking, gooseberries and currants need to be separated from the stems and the end blossoms. For small currants on a strand, a fork makes for a quick cleaning. Simply pull the strand through the fork tines, and the berries detach easily. Gooseberries are berry by berry cleaning. It's much like pitting cherries—you just sit down and do it.

# Storage

Both currants and gooseberries have the added benefit of keeping well in the refrigerator until you are ready to use them! We have been able to store them for up to 4 weeks without deterioration of flavor or texture. Make sure the humidity levels are moderate—not too humid, not too dry.

The berries should be stored uncovered in the refrigerator, with good air flow. Between 32 and 36 F is ideal for keeping the berries for an extended time. The berries will continue to ripen, even in cold storage. As they ripen, the berries start to soften. Soft berries are much harder to handle. It is always best to get the berries to the end user as soon as possible.

It is best not to store them near apples, because of apples' natural propensity to exude ethylene. Fortunately, apples are not usually in season when it's time to harvest currants or gooseberries. Be sure to remove any over-ripe fruit from the vicinity, as it will also give off natural ethylene, which hastens the ripening process.

## How to sell

These *Ribes* are less familiar to the public than blackberries and raspberries. I did a lot of sampling to customers the first year we had them. After that, they sell themselves. That being said, they are popular among people who like to make their own jellies, wines, and specialty desserts. If part of your business is making jelly or wine, then these are good candidates. We sell the fruit fresh and let people make what they wish.

As always, I eat everything fresh and don't worry about turning it into something else. The fruit is a bit tart (not as tart as rhubarb). I have convinced some people to snack on these, but most people who get these fruits turn them into pie, sauces, or jellies and jam. By the way, gooseberry jam is delicious, and red currant jelly is beautiful as well as delicious. If you make jam or have someone you partner with, gooseberry jam is a great unusual product to add.

How you sell is going to depend on your sales model. Do you sell to individuals, to restaurants and chefs, to wineries, to specialty stores, or to a wholesaler?

If you are selling to individuals, start with a few bushes and start sampling them. If you sell at farmers' markets, currants and gooseberries act as a marketing ploy. It's a place to start a conversation—what is it, what does it taste like, what do you do with them. Samples are helpful! Even if they don't buy the berries, they are most likely to buy something else. Plus, the people who know what they are will be thrilled. Their excitement will prompt others to try the berries as well.

If you sell to restaurants and chefs, you can approach them first before you plant. Find out if this is something they are interested in. If you have two different chefs who are interested, others will be interested as well. You will be able to sell easily in your area. Chefs at high-end restaurants will be interested in gooseberries and currants. It helps the chef to let them know when you expect to have them. Good chefs are planning their menus 3-6 months in advance. Restaurants that specialize in local and fresh produce will often have daily or weekly specials into which they can incorporate unusual fruit.

Get an idea of the quantity that they might want, which will help you with determining how many plants you want to put in.

If there is a winery or brewery in your area that produces specialty drinks, I can assure you they will have a great interest in these products. Talk with them, and see what kind of quantity they need to make a batch. Because you are selling in larger quantities, perhaps your whole crop to one buyer, you will have to determine what an appropriate price is. If you would sell the berries for $10/pint in a retail setting, what would you sell 100 pounds for? $5/pound would be a starting point. The wholesale price of raspberries could be a consideration, but you will probably sell for less than that if you sell in bulk to one buyer.

If you sell to specialty markets, then you are looking at wholesale prices. Since currants and gooseberries rarely show up on the USDA wholesale pricelists, you will determine the price. Again, look at the going rate for raspberries and price accordingly. You will need to have your berries in the appropriate sales container, such as a pint or half-pint. Ventilated clamshells are ideal for markets because the berries do not spill out. They also help keep people from touching every single berry in the basket.

You can get up to 10 pounds of fruit from each mature gooseberry plant (3 years old). You can harvest 3-8 pounds from a black currant plant, and 1-3 pounds from red or champagne currant plants. This gives you an idea of how much you might have for yourself or to sell. Because the black currant plant is larger, it produces more, the berries are larger and easier to harvest, and the flavor is more intense. Black currants are popular for sauces, jellies, and wines. We sell red and champagne currants, mostly to restaurant chefs and home chefs.

Because no one else in our area sells gooseberries or currants, we can sell these for a premium price - the same price as raspberries. We sell all of our currants and raspberries fresh to individual customers at retail prices.

Gooseberries and currants rarely show up in the wholesale produce markets, so do not be afraid to price them as an exotic fruit. Although they are much easier to grow and store longer than raspberries, they are much harder to come by. The scarcity value keeps the price at a good level.

## Propagate your own

You can divide clumps of gooseberries every five to ten years, or so, as a way to increase your own holdings. Simply dig up the entire clump, separate out into four or more sections, each with roots, and re-plant about four feet apart. We took our initial three gooseberry plants and made two rows of about twenty plants. Because we do not have the plants certified mildew resistant, we will not sell any of the stock.

Another way to propagate your own gooseberry plants is "tipping." You can bury the tip of a cane, still attached to the mother plant, in the soil and weigh it with a rock. It will take about a year to grow strong roots, so put some stakes around it to protect it from weed whackers, people, or animals knocking it loose. After a year, you can clip it to separate it from the mother plant and then transplant it to its permanent location.

We find that enough canes grow on their own and the plants multiply without us having to do anything. Every 4-5 years, we separate the bushes and re-plant the new stock.

Since most gooseberries are hybrids, trying to grow from seed is not a good plan. You could try it as an experiment, but there is no telling what will grow. You are much better off "tipping" a branch or dividing it at the roots.

If you want to propagate gooseberries from cuttings, you will do so when the plant is dormant, just like currants. Cut from last year's growth, with 8-10 buds on the cutting. I find the cuttings method easier than tipping, although the plants will tip-root on their own at times. Easiest of all is to divide the mother plant.

Currants can be propagated from cuttings. When the plant has lost all its leaves and is dormant, take a cutting from a cane of about eight inches. You want about five or six buds on the cane. Plant the cutting so that at least two buds are in the soil, about half the cutting. You can plant in the ground, extending your currant row, or plant into pots. Firm the ground around the stem, and water well. If you plant into pots, remember that the currants want all those cold days in order to set up for the next growing season. Leave in a cold area where it will stay below 45 F. As with gooseberries, most American currants are hybrids. Grow from cuttings rather than from seed. The seeds will not grow to be the same as the parent plant.

You can either plant your currant cuttings into the soil or into pots. I find it's easier to keep the roots moist when they are in pots. You will see new shoots the following growing season. By then, the new plant will have a good root system. Then plant in your designated area.

In general, there is about a 50% success rate for gooseberry and currant cuttings planted in the ground and about a 75% success rate for cuttings in pots. Because of the time involvement and the moderate success rate, we choose to divide the plants or purchase new stock.

# Figs

Figs, *Ficus caricus*, like hot weather, the hotter, the better! As natives of the Middle East, consistent hot weather is their preference.

Fig trees followed the routes of exploration from the Middle East, spreading throughout Greece (Aristotle and Theophrastus both weighed in on figs) to Turkey (the world's largest producer of figs today), all of the Roman Empire, Afghanistan, Portugal, India, China, and England. And the Spanish first brought figs to California in 1520. Not only did people take fig trees with them to plant in the newly conquered or trading areas, but they were also compelled to tell everyone that they planted more figs in the new country.

The Italians in the 19th century get much of the credit for bringing many varieties of figs to the United States, as they brought their beloved trees when they immigrated. We were the beneficiary of this trend.

Once upon a time, a long time ago, back when there were classified ads in the newspaper and people read those ads, my mother saw an ad for fig trees in Wilmington. "Hmmm," Elaine thought. "If he can grow figs in Wilmington, so can I." Definitely a competitive streak in the family.

Elaine and my youngest brother drove to Wilmington to find this gentleman (using a map and my mother's knowledge of Wilmington) who had fig trees growing in his backyard. "Here they are," he said. "Which one would you like?" They picked out a tree, dug it up, and brought it back. This happened in 1978 or so. This fig tree is still with us, a variety called Celeste, small, very sweet, and delicious.

An older Italian gentleman saw my mother's precious figs in the farm market; "Did you grow these?" "Yes, I did," was the proud answer. "Well, I have figs that are much larger and very good." A brief moment as they eyed each other. They agreed to trade cuttings. And that is how our second tree came to us, about 1982. That variety is Brown Turkey, still with us and growing well. Apparently, fig trees can live to be 100 years old or so; we will find out.

This just goes to show that we are not immune to that urge to grow something just because we want to. There definitely was no thought about how profitable growing figs might be. We were fortunate from the beginning that our customers love figs as much as my mother does.

The challenge of figs is cold winter weather. Figs need winter protection from cold and wind if your winters get below 25 F. Figs need a minimum of 6-8 hours of direct sun, and prefer sun up to sun down for direct sunlight. Figs can grow 10-20 feet tall and 10-20 feet wise. They are commonly called trees but typically have more of a bush shape. You can prune a fig so that it has a central trunk and then branches that are higher off the ground. In our desire to limit our workload, we leave the fig trees to grow branches from the ground, which creates a round, bush shape. This makes it easier to dig roots for propagating more trees. Figs are grown in zones 5-10, with varying degrees of difficulty.

What we call the fruit, the actual fig, is a syconium. The fig is an enclosure, called a syconium, with hundreds of little flowers inside that develop into hundreds of mini fruits (drupes).

This is a complicated way to describe something that is delicious. For our convenience, we are going to call them figs and fruit and not worry about nomenclature.

After a lot of trial and error, we have finally come up with a system that means we get an abundant fig crop each year. We are in growing zone 7a, so we have winters regularly dip to 10 F and occasionally to zero. We have snow, wind, ice, and sleet. Winters can be very hard on fig trees. We are usually done with frost by late April, but our official frost-free date is May 10. We can have frost as early as October 10, but usually not until November. However, we start getting cool nights in September and downright cold in October. Growing figs is all about night-time temperatures.

This is a perfect example of growing something because we wanted to see if we could do it, not because there was a business model for it. We were fortunate that figs are in high demand and that figs have become an extremely profitable crop for us. As we learned about figs and saw how easily they sold, we became more organized in how we approached growing them.

I'm not saying you should stay away from growing something just to see if you can. Simply recognize that is what you are doing. Growing for profit requires being deliberate in your choices.

# Varieties

There are hundreds of varieties of figs! These are a few that I am familiar with and these can grow in a variety of climates. You can see some of the varieties show up in several different categories—versatility is a wonderful thing in a plant! As you look at different varieties, check the growing preferences and winter hardiness to see what will suit your particular area the best.

Many varieties of figs will produce two crops each year. The first early crop is called "breba," and the second crop, later in the summer, is the main crop. An early breba crop is a bonus, although it is usually about one-third the quantity of the main crop.

Figs have been cultivated and prized for thousands of years. Delicious, nutritious, and beautiful. Immigrants brought small trees from their homeland to grow in their new homes. As a result, every area has its own favorite varieties.

We have settled on Chicago Hardy (because we need something hardy for our winters, prolific and flavorful), Brown Turkey (my mother's second fig tree, large, sweet, and hardy for our area), and Celeste (a small brown fig, my mother's first fig tree, absolutely delicious, and also very hardy).

The Chicago Hardy fig is more winter hardy than most varieties and is a relatively sturdy fig for the retail business; that is the majority of what we grow. Black Mission figs are the fig of choice for commercial growers and for drying. They have a more durable skin and are not as juicy as other varieties, which means they do well for shipping and dehydration.

<ins>Why we do not grow white figs</ins>

White figs are delicious, but perilously delicate to handle. The skin mars easily and does not handle well. For home growing, they are fine, but for profitable farming, they are a lot of work. Their skin is much more fragile than the darker figs. Every time a white fig touches another fig or the container or a finger touches it, there will be a dark mark on the skin. The skin shows everything. They are much more difficult to handle if you are selling them. Since people buy with their eyes, there is resistance to buying the white figs. We choose to grow darker figs which are easier to sell.

Many varieties of figs will have an early crop, called a breba crop, if the weather is cooperative. These figs grow on the previous years' branches. The main crop grows on the new growth. In colder zones, it is more difficult to get the breba crop because of how late the branches come out of dormancy. We find that we will get a very light breba crop in June, but the bulk of the figs come in September.

**For hot dry climates:** Penache Tiger Stripe, Black Mission, Brown Turkey, Peter's Honey. Of these Black Mission is the variety most commonly grown for commercial production, as it has a sturdier skin and will ship reasonably well. Just for that reason, growing a different variety of varieties helps set you apart from a large operation.

**For cooler coastal climates, such as the Pacific Northwest:** Desert King, White Genoa, Olympian

**For cold-hardy varieties:** Chicago Hardy, Celeste, Brown Turkey, Desert King

**For humid climates:** LSU (Louisiana State University) Gold or Purple, Celeste, Kadota, Peter's Honey

**Container varieties:** Consider Chicago Hardy, Violette de Bourdeaux, Peter's Honey, LSU Gold or LSU Purple, or Corky's Honey Delight. These are semi-dwarf varieties that will handle container growing better than large-growing varieties.

## *The varieties:*

**Black Mission**: grows well in zones 7-10, cold hardy to brief periods of 10-15 F. The figs are medium-sized, dark nearly black-skinned fruit with red flesh and a sweet flavor. It is the most popular variety grown in California.

**Brown Turkey**: grows well in zones 7-10, cold hardy to brief periods of 0-5 F. The figs are medium to large with light brownish rusty colored skin, and light pink flesh inside. Sweet flavor reminiscent of melons and honey, very juicy. Will more frequently have a breba crop than other varieties. They will split at the bottom if there is a lot of rain when the figs are close to maturity.

**Celeste**: grows well in zones 6-10 and are quite cold hardy. Small to medium-sized fruit, with a bronzed purplish skin, and reddish/pink flesh. One of the sweetest figs, with a rich flavor. The skin is quite tender when fully ripe. We rarely have a breba crop with this variety. However, the main crop starts earlier than other varieties (smaller fruit, shorter time to maturity), and the trees are remarkably prolific. Very resistant to pests and rarely split due to excess water.

**Penache Tiger Stripe:** grows well only in southern zones 8-10. Hot summers, warm winters, not very cold hardy. Generally, a smaller tree, about 12-15 feet, this variety has a striking green and pale yellow striped fruit with dark red flesh. Excellent flavor, with a late ripening main crop. Because this variety takes a long time to reach maturity, there is only the main crop.

**Desert King** (sometimes just called "King"): a versatile variety that grows well in zones 6-10, with cold hardiness to 5 F. It grows quickly and can reach 15-25 feet easily. One of the "white" figs, the large fruit is a pale green color at maturity with pink flesh. This variety thrives in slightly cooler temperatures, making it a favorite for areas with moderate summers. It has an unusually large breba crop and a smaller main crop.

**Violette de Bourdeax**: a semi-dwarf variety, zones 7-9, and reasonably cold hardy to brief periods of 5 F. It has a small purplish black fruit and red flesh. Excellent sweet flavor, almost berry like. These produce both good breba and main crops and prefers lower humidity.

**Chicago Hardy**: as you might guess from its name, this is one of the most hardy varieties of figs, suitable for growing zones 5-10. The roots will survive temperatures to -20 F! The branches will survive to about zero degrees so winter protection is desirable if that is the kind of winter you have. The fruits are medium-sized with a burgundy purple skin and light pink flesh. The flavor is sweet and berry like. Chicago Hardy are heavy croppers - each tree can produce 100 figs per season. Usually, there is a light breba crop and a heavy main crop. This variety is well suited to growing in a bush shape and maintaining a height of around 10-12 feet. In addition to being extra hardy for cold winters, we find that the fig trees produce longer during the fall than any other variety. The trees hold onto their leaves longer and more figs ripen late in October than other varieties.

**Corky's Honey Delight**: a semi-dwarf tree suited to zones 7-10 and the roots will survive dips to 5 F. The medium to large-sized fruit is pale green at maturity with dark pink flesh. Sweet, almost honey-like flavor, this variety has both a strong breba and main crop each year. Good for container growing.

**Peter's Honey**: good for zones 6-10, with root hardiness to 5 F. Native to Sicily, this variety prefers areas with warm summers. Peter's Honey produces a strong breba crop as well as a good main crop. The figs are a pale yellowish green at maturity with light amber colored flesh. Very sweet, with a syrup-like flavor. In common with all the "white" figs, it has tender skin. It is a container-friendly variety.

**LSU Gold and Purple**: developed by Louisiana State University, these varieties are good for the southern United States, and growing zones 8-11. They are both container friendly, with medium sized fruit and sweet flavors. LSU Gold is a white fig, and LSU Purple has a medium reddish/purplish skin. They are prolific, small trees, and grow well in areas with warm winters. Well suited for container growing.

**Olympian**: good for zones 6-10, with a little extra winter hardiness. It is a compact variety that has unusual green and purple striped skin with dark red flesh. Very large fruit, with tender skin, so beware handling. This can provide a good breba crop if winter temperatures stay in the teens or higher, as well as a strong main crop in late summer. Exceptionally sweet!

There are lots of varieties to consider. Look at winter hardiness, how prolific, summer growing preferences, and humidity preferences, to find the right varieties for your farm.

# How to grow

**Planting**

Each tree can get quite large. We prune our figs to develop width rather than height; this makes it easier for picking. Plant your figs at least 12-15 feet apart.

Dig the hole about twice as wide as the pots or the root ball and a little bit deeper. Cover the roots but do not bury the trunk. Amend the planting hole with some quality potting soil or compost, but you do not need any additional fertilizer at this time. Water well to help the roots establish. Add a layer of mulch, leaving a 3-inch ring immediately around the trunk clear of mulch.

Figs that grow on the older branches are called "breba" figs, and these will ripen in late spring/early summer. The main crop develops on the new growth, and they ripen in the late summer/early fall.

If you have pests such as gophers, rabbits, or deer that like to nibble on young bark, then you may need to put a wire cage around the young trunk. If you have gophers, then you will sink the wire cage into the ground to protect the roots. The wire cage should be at least two feet in diameter and about 12 inches deep in the soil. Use wire that is a half-inch in pattern to keep gophers or other rodents away from the roots. The roots will be able to grow through the wire and the main roots will be protected.

Check to see which varieties are suitable for your climate, whether hot summers or moderate, extra cold winters or warm winters, and high humidity or more moderate humidity levels. Also, look at the tree size. Many of the figs can be 15-25 tall. To keep the fig within easy harvesting height, you will do some pruning each spring. Keeping the tree a bit shorter will cause it to grow wider, so be sure to have enough space for lateral growth.

Fortunately, the branches are quite flexible, so it is easy to bring a branch down to you when you are harvesting.

Since figs are self-fertile, you only need one to get figs. However, they are so delicious and they sell easily, so consider space for more. Fig trees will often produce a few figs in the second year, and more each year following.

Many fig varieties will survive winters to 5 F, but then they die back to the ground and start all new growth. This means the roots survive, and that the plant needs to grow all new top growth. By the time figs appear, it may be too late to harvest the crop. It's very depressing to look at figs on the branches that are not ripe and will not ripen, because it is now too cold. Figs tend not to ripen if the nighttime temperatures go below 50 F. We always hope for nighttime temperatures to continue in the moderate range for as long as possible.

To counteract that, we want the fig trees to have a head start in the spring. We decided to expand the number of fig trees that we grow because the demand for figs is so high. Now, we have most of our fig trees planted in an unheated tunnel and then cover them with a row cover as well. This gives them a 2-3 week head start in the spring. After nighttime temperatures start hitting 50 F consistently, we remove the row cover and plastic tunnel cover. I prefer to leave the greenhouse plastic on until the fig trees are pushing against it. The fig trees easily will grow 5-6 feet each year, well above the top of the tunnel.

You can plant the tree in the ell of a house or building so that it gains some of the building's residual warmth as well as protection from winter winds. This is helpful, but we found that we still had years when the fig trees did not have enough time to ripen many figs. The double layer of protection works well for us.

I know farmers in zones colder than us who are growing figs successfully. The colder your winter, the more protection you need to provide.

If you are in warmer growing zones, your figs will be very happy. Sunshine and heat are the two qualities that will bring you an abundant crop. If you are in a cooler growing zone, then it's up to you to provide that additional protection so the figs can fulfill their mission.

**Winter protection**

Multiple layers do wonders for figs in the winter. There are many ways to achieve this. You can build a mini-greenhouse with stakes around the tree. You can use twine to tie the branches closer together, and then fill the branches with leaves. Then wrap burlap or row cover or another fabric around the stakes. These are breathable layers that allow air to reach the branches and water to reach the roots.

Ideally, you do not want the fabric to touch the branches, because if the fabric gets wet it can freeze and get stuck to the branches. That can cause damage to the branches. In the spring, uncover the outer layer as late as possible. You can push the leaves out and leave them around the tree to cover up the roots. The leaves can be the mulch for the new year.

Another option is to trench your fig tree. Fig branches are fairly flexible. Dig a trench next to the tree. Pull the branches to the side and to the ground. Cover the whole tree with soil and leaves. Essentially, you are putting the whole tree in the ground. In the spring, you will dig the tree out and let the branches move upright again. If there is still a danger of frost, then use row cover to cover your tree until it warms up.

If you are in zones 5 or 6, I recommend using three layers, such as leaf mulch, row covers, burlap, trenching, or an unheated tunnel.

Do you have to build a tunnel? Absolutely not! We did that only when we decided to expand our fig production. Prior to that, we had (and still have) fig trees in the ell of a building, taking advantage of the warmth from the building. Mulch and row covers are staples for us. Our 50-year-old fig tree is not in the tunnel and is still quite prolific.

Fig trees are self-fertile, so you only need one. We prune the trees back to about four feet tall in the late fall. You can use a double layer or row cover or build a box around the tree and cover that. In addition to the cold temperatures, the fig trees react to the wind. Winter winds will dehydrate the tree and cause more branches to die off. Preserving the older branches is key to getting the early crop of figs.

**Fertilizer**

A good top dressing of aged manure or compost applied each spring is helpful for the tree to produce figs. We will add compost in the middle of the season as well. This helps the size of the figs.

**Water**

Water a new tree weekly for the first summer. After that, the tree should be sufficiently established. Usually, the fig trees do fine on their own. However, if it is an unusually hot and dry summer, with no rain for a month, then apply a long watering once each week. Figs depend on consistent watering in order to produce abundant fruit.

**Pruning**

In the early spring, thin out any dead branches. We also clip out small branches. Air circulation is key for plant health.

For the fall: If you cover your fig trees for the winter, prune back to the desired height after the tree has dropped its leaves. We trim ours to 4-5 feet tall so that they fit in the tunnel. Remember, as you top the trees to the desired height, it will cause the tree to branch out to the side. If there are any sickly-looking branches, remove them so that the tree focuses its energy on good, healthy branches and fruit.

Do you need to do any additional pruning? No.

Can you do additional pruning? You can clip side shoots or small branches.

Will pruning side shoots increase how many figs you get? No.

From our horizon, additional pruning is something that takes more time but does not yield more fruit. Therefore, we do not do it.

Spring is also the time to dig up any new fig trees that have multiplied at the root level. You want to do this while the plant is still dormant for the best success. Separate with a shovel and then plant into a 2-gallon pot. Keep it well watered and be patient. The fig tree will work on growing roots. It may take a month before you see any leaves growing.

Fig trees will start to produce figs after one year, and produce a full crop after 2-3 years. As the tree gets larger, you will get more figs.

## Figs in containers

You can grow figs in containers—the larger the container the better. The trees will not be as large nor produce as many figs as when they are planted in the ground.

If you are pressed for sunny space, though, this can be a good way to have figs. In addition, you can move the container to a protected area such as a shed or garage for the winter. In a container, you will need to water nearly every day. In the container, count it as one to two growing zones colder, since the roots are now above ground.

I recommend wrapping your fig with a protective layer even if you move it into a garage or shed unless that area will stay above freezing all winter. When you move the tree back outside in the spring, a layer of row cover will continue to protect the new growth until the spring weather is consistently warm.

You will need to feed your figs with a top dressing of compost every month to give it enough nutrients to produce figs. Producing fruit for any bush or tree is a lot of work, so it is important to keep the tree well-nourished.

You can have multiple containers as a way to increase your fig crop. In general, if you want to grow figs as a profitable endeavor, I recommend planting them in the ground. We have had figs in containers, and they produce 10-20% of what the tree will produce when planted in the ground.

# Maintenance

Once your fig tree is established, there is very little you need to do.

In the spring, remove winter protection layers, slowly. If you are in a colder area, leave a covering as long as possible. It will not get too hot for your fig tree. Add a top dressing of compost or aged manure around the tree. Add mulch. If you did not prune in the late fall for winter covering, do your pruning now. Remove any dead branches. You can save the clippings for propagating new trees.

If you have a prolonged period of no rain, then you may need to water. We only had to do this when we went for 8 weeks without any rain.

If your fig trees are in containers, then you will need to water nearly every day. You will also want to add some fertilizer about once a month. A granular slow release fertilizer works, or a top dressing of compost.

Check for an early crop of breba figs. In the late summer, you will start to harvest the main crop of figs.

After the crop is finished and the leaves drop, prepare your trees for winter. Cut back if needed. Again, you can save the cuttings to propagate more trees. Use as many layers of protection as needed for your area.

# How to harvest

Of our five fruits, figs are the only fruit that does not ripen off the tree, so it is important to harvest when they are fully ripe—but not too ripe! When harvested properly, they will keep a week or longer in the refrigerator. The skin is very tender and will tear easily. Hold by the stem and place gently into the container.

It is important to harvest into a container that is "soft," such as the green pulp quart boxes. A wooden box or woven basket can tear the tender fig skin. If you have hard containers, line them with parchment paper so that the fig skin is not bruised or torn.

It is ideal if you can harvest into the final container so that the figs are handled as little as possible.

You do not want to go around squeezing your figs to see if they are ready. Instead, you are going to look for a change in the way the fig is on the branch. As it is growing, the fig sticks straight out or up. The stem is perfectly straight. As the fig ripens and turns from green to tan to brown (or from dark green to light green), the stem will soften and the fig will droop downwards. If the fig does not come off easily with a pinch of the fingers, it is not ready. Leave it for tomorrow!

If figs are nearly ripe and you get a lot of rain, the figs may split at the bottom. The larger figs are more prone to this than the smaller fig varieties. If the fig has a split, then you will want to harvest and refrigerate immediately. Either freeze or use within 24 hours, as the figs will start to sour quickly if they are split.

Our main harvest month for figs is September. If we are lucky, we have an early crop in June (all dependent on winter and spring weather). Most of the time, it is late August through the first week of October. Last year, we had figs through the whole month of October. A warm fall kept the figs going. As the nights cool, the ripening slows dramatically. In a warmer zone, you could have figs from June through October or later.

Be careful of the white sap from the stem or new growth on the trees. It can be an irritant for your skin.

Figs will continue to ripen on your tree for as long as the weather is favorable. When the night-time temperatures drop into the 50s, then the ripening slows dramatically. Instead of figs ripening every day, it will take several days to get from almost ripe to fully mature.

# Storage

If you have harvested at the correct time, when the figs are only slightly soft, you can store them in the refrigerator or other cold storage for 7-10 days.

The ideal storage temperature is 33 F. Do not store near apples, because of the ethylene that apples naturally produce. That will reduce the shelf life of your figs.

Store uncovered in soft cardboard containers. Anything that is hard or has sharp edges can damage the skin of the figs.

The figs will continue to soften while in cold storage but will not sweeten any further.

The figs can be two or three layers deep in this condition. If the figs are very soft and/or the skin is slightly wrinkled, then store the figs in a single layer. As the figs get soft, the skin becomes very tender and will tear easily. The figs are very difficult to sell to retail customers at this stage.

Extra soft figs may still be of interest to chefs who are cooking with them. Your options will be to freeze them, make jam, dehydrate them, or sell them at a discount to someone who can create a fig syrup or other condiment.

# How to sell

The good news is that figs sell themselves, unless you are in an area where everyone has a fig tree in their back yard. In which case, you might want to grow something else!

How you sell is going to depend on your sales model. Do you sell to individuals, to restaurants and chefs, to wineries, to specialty stores, or to a wholesaler? Do you have a farm stand, go to farmers' markets, or set up in a business food court?

If you are selling to individuals, start with a few bushes, maybe even just one. If people in your area have not tried fresh figs before, then you can sample some. If you sell at farmers' markets, fresh figs act as a marketing ploy as well as something to sell. It's a place to start a conversation—what are they, what do they taste like, what do you do with them. Samples are helpful! I've never had someone try a fig and then not buy some. The sweetness is amazing, without any added sugar. You do need to warn people that fresh figs do not taste like the soft cookie with fig in its name. In addition to trying a sample and purchasing a cup of figs, customers most likely will buy something else. Plus, the people who know what fresh figs are will be thrilled. Their excitement will prompt others to try the figs as well.

I had people rushing to the farmers' market on Saturday mornings so they could be sure to get their figs. Then they also bought lettuce and broccoli and onions, etc. It was like having a neon arrow pointing to our tent.

If you sell to restaurants and chefs, you can approach them first before you plant. Find out if this is something they are interested in. Find out if they prefer larger or smaller figs and how they might use them. See if they will do any cooking with figs and if they would be interested in any very ripe figs. In spite of our best efforts, it is all too easy to end up with some figs that are too soft for sale at a farmers' market but are ideal for cooking.

Be bold in approaching chefs about figs. Any chef who creates signature dishes will be very interested in fresh figs. Let the chefs know well in advance that you expect a crop in late summer. Many chefs plan their menus 3-6 months in advance. Look for restaurants that have daily specials or who focus on using local produce. These chefs will be happy to support you by purchasing your figs as well as other produce that you grow.

Get an idea of the quantity that they might want, which will help you with determining how many plants you want to put in.

It is possible that someone making wine, "shrubs," or cordials could use figs, but a more likely option is fig preserves.

The benefit of a jam or preserve is the long shelf life. This also gives you a way to sell figs when they are not in season. Check with your state or county regarding regulations for cottage industries for farms. Some states' Department of Agriculture will serve as the inspectors and regulators. We choose to send produce to a family which has an

FDA-inspected operation. Having fig jam does not mean that you have to do the work.

You can also freeze figs easily. Cut the stem off, cut the figs in half, lay them on a tray, and freeze. This gives you time to decide if you are making preserve, dehydrating, or turning into something else. The figs will be soft when they defrost, so cooking or dehydrating are the options.

If you have a dehydrator, then dried figs are another way to preserve your harvest to sell at a later time. Their natural sweetness is intensified in the dried state, so it is not necessary to add any sugar.

If you sell to specialty markets, then you are looking at wholesale prices. Be sure to check the wholesale price of figs to find a comparable price. Unfortunately, figs rarely show up on the USDA wholesale pricelists, except perhaps in California and Arizona. You may have to determine the price on your own. You can look at the going rate for raspberries as a good starting point. Raspberries are a comparable premium fruit. We often use the price of raspberries as a guideline for how to price unusual fruit.

Do not be afraid to charge a retail price based on the going wholesale price. Your product may not be traveling across the country, but you are offering a fresh, high-quality product that deserves a good price. Figs often sell for more than $1 each.

You will need to have your figs in the appropriate sales container, such as a pint or half-pint. Ventilated clamshells are not good for figs, as the sides are hard on the tender skin. Frequently we see the standard pint or half-pint green cardboard containers with a small "hairnet" over the top to keep figs from falling out, or to keep people from switching figs or removing them from the display. Be sure to include the cost of the containers in your pricing.

At farmers' markets, figs will draw people to your table. Few people grow figs, so these are a marketing draw.

Just a comment here: when you sell to chefs and restaurants, don't feel you have to discount the price unless they are getting a very large quantity. You are most likely delivering, plus providing an unusual product that they can't get elsewhere.

Find your market report at the USDA MyMarketNews website:
https://mymarketnews.ams.usda.gov/filerepo/reports

I use Philadelphia's:
https://www.ams.usda.gov/mnreports/na_fv020.txt

# How to propagate

As farmers, growing more plants is what we like to do. If something is going reasonably well, we try to figure out how to have more plants without spending any more money.

Figs propagate fairly easily.

You can cut during the growing season. Select a branch that is growing horizontally and does not have any figs on it. Cut a section about 12-15 inches long. You want to see a leaf bud at the end of the branch. Remove all the leaves, leaving the small leaf bud. Score the sides with a sharp knife to remove the bark for about 4 inches off the bottom, up to the first leaf node where you clipped the leaf off. Use a 10-inch "tree tube," which is a tall, slender pot. This gives the plant room to establish roots in a narrow space.

Us a mix of 60% potting mix, 20% compost, and 20% coconut coir or peat moss. The coconut coir/peat moss helps retain moisture in the soil. Insert the fig cutting into the soil in the tube. You will only leave about 3-4 inches of the stem above the soil. Press the soil tightly around the stem. Water.

Now place a stick that is taller than your fig cutting next to it in the pot. Drape a piece of plastic over the cutting and the pot. Tie a string around the pot to hold the plastic in place. You have created a mini-greenhouse. You will leave this on for one week only. The moisture from the soil is trapped inside the mini plastic greenhouse. After you remove the plastic, water the soil weekly to keep it slightly moist.

It will take about 3 months for the new tree to have a decent root system. Wait until spring to transplant into the ground.

Using a shorter pot will decrease the success rate. If you do not have a tree tube, then use a one-gallon pot. You want the depth so that most of the cutting is below the soil.

When you prune in the fall, to cut back to the height that you want, you can use the pruned branches to root them. Make the cutting about 10-12 inches long, and cut below a node. Since the leaves have already dropped off, no worries. You can use a rooting hormone to help promote root growth. Dip the cutting into the rooting hormone and then place in the damp soil. Have at least half the cutting in the soil. You can use the rooting hormone for prunings or for the cuttings.

If you have started several plants, you can take a clear storage tote and turn it upside down to create your mini greenhouse.

We have tried doing cuttings, and it works.

However, if you are a busy farmer, and want to minimize your work, here is how we now do it. We wait till the tree is 2-3 years old. We dig up some of the roots of the plant in the spring. The root system will have spread out, and it is easy to cut a chunk off with the shovel. We then pot up the divided pieces into gallon (or larger) pots and leave them in the germination house. Any area protected from wind and cold will do. Since there are some roots with the stalks, we don't have to create mini-greenhouses. Because we have the pots where we are starting seeds, it's easy to water as needed. There will not be a lot of top growth for the first month, as the figs are busy building a good root system. It takes about 2 months for the figs to look like something that is growing. This way, we don't have to do any maintenance on small pots over the winter.

You can sell the young trees or plant them to increase your fig production.

# Fig wasps

Have you heard of fig wasps? Does it sound weird? It is unusual how figs are pollinated.

Most pollinators fly from flower to flower, transferring pollen as they go. Fig wasps travel from flower to flower as well, but all the flowers are inside a single fig.

Remember, we talked briefly about figs being syconium and the flowers are actually on the inside. A tiny fig wasp (*Blastophaga psenes*) actually crawls inside the fig with pollen from its birth fig and then pollinates the flowers and proceeds to die. The fig releases an enzyme that dissolves the wasp. You will not see any remains of the wasp when you eat a fig. Any crunch is from the fig seeds. The wasps are about 1.5 mm, or .06 inch, in length, so very tiny. You won't see them.

The fig wasp's life span is about 2-3 days. A newborn female wasp will exit the fig at the eye (opposite the stem), gathering pollen as she goes from the interior florets, and then seek out a new fig to fertilize. The male fig wasp is born, seeks out a female within the fig, and then dies, never leaving the fig. Enzymes go to work.

If you have figs that fell off the tree at the immature stage, they were not fertilized.

The fig trees need the wasps to pollinate, and the fig wasps only fertilize figs. It is a very specific relationship.

# Pawpaws

Pawpaws, *Asimina triloba*, are a tree fruit, and, in common with our theme, they need no sprays or special treatment. They have large oval leaves, and the fruit itself is vaguely tropical in appearance, a bit like a mango. They come by this tropical appearance honestly, being part of the family Annonaceae. All other members of this family are subtropical or tropical plants. This is the only one to be able to tolerate cold weather.

Pawpaws are native to North America. In the past fifteen years, pawpaws have developed a dedicated following for this fruit which is sometimes called the American banana. It grows well in a variety of growing zones, from 4-9. Pawpaws are cold hardy to -20 F. Your pawpaws need at least 2 different plants for best fertilization. The earliest written record of the North American pawpaw is from the Spanish explorer, Hernando de Soto, in 1541. He was introduced to the fruit by the indigenous people near the Mississippi River. Native people and European settlers depended on this fruit for centuries. Lewis and Clark documented the value of finding pawpaws to eat on their journey.

This is North America's largest tree fruit native to the continent and is indigenous to 26 states, from Nebraska to Florida and as far north as Ontario. It is also the state fruit of Missouri.

After being one of the most popular fruits in the United States in the early 1900s, pawpaws fell out of favor after World War II. The rise of a global and standardized food system had no room for a short-season odd-looking fruit. Fortunately, there has been a resurgence of interest and appreciation for pawpaws in the past two decades. There are now pawpaw festivals in many states, providing inspiration to try fresh pawpaws and pawpaw-inspired recipes.

The skin is a light green/yellow with dark speckles. The flavor is a cross between a banana and a mango, with a soft, custard-like texture when fully ripe. The size can vary from a quarter pound to a pound in weight. The seeds and skin are not edible at all. Pawpaws typically ripen in late August or early September. In some areas, certain varieties can mature in October. The fruit on each tree will ripen over the course of about two weeks. It's a short season!

Do not confuse these with papayas, *Carica papaya,* a true tropical fruit that is also called pawpaw or papaw in the countries where they are grown.

The mature pawpaw tree is about 25 feet tall, and about 15 feet wide. The tree will grow suckers—new trees—from the root for the first four to five years and then stop. It is best to prune the suckers off so that the tree is not competing for sun or nutrients. For best fruiting production, two different varieties are desirable.

Native pawpaw patches can be found along rivers, streams, and ravines. There is a pawpaw patch along the Potomac River just north of Washington, D.C., and another about 10 miles south of the capitol. Often found in wooded areas, the trees are strong survivors. However, fruiting will be sparse unless it receives a lot of sunshine.

# Varieties

There are now many cultivated varieties as well as the native *Asimina triloba.* The cultivated varieties have larger fruit and a larger proportion of flesh to seeds. The native pawpaw can be almost half seeds. While the seeds are dramatic in appearance, more fruit to eat is better if you are looking for consistent sales. The cultivated pawpaw will be about 90% flesh or more. Since you will have better pollination and cropping with two trees, preferably different varieties, be sure to add a cultivated pawpaw variety to your mix.

Although native pawpaws are found throughout the eastern half of the United States and Canada, the cultivars are quite adaptable to growing conditions throughout the whole North American part of the continent.

You can extend your pawpaw season by planting varieties that mature at different times. However, you generally need two trees for pollination, so the flowers need to be open at the same time. Have two trees of early ripening, two of mid-season, and two of late-season ripening for best results.

There are many varieties! Look for winter hardiness, how prolific the crop is, if the variety is susceptible to cracking, how large the fruit is, and if it is early, mid-, or late-season ripening. Find the varieties that will work best for you.

**Varieties for zones 4-5**

Look for early ripening varieties such as Allegheny, Summer Delight, and Pennsylvania Golden. It takes about three months from bloom to maturity for early ripening varieties, compared to 3.5 to 4 months for others.

**Zones 6-9**

Any variety will work well. If you are in a semi-arid climate, you will need to provide 3-5 inches of water per month. Allegheny, Collin's Select, Eastern Seedling, Halvin, KSU Atwood, Maria's Joy, Shenandoah

**Allegheny**: growing zones 4-9, ripens early, 8% seed, which means only 8% of the interior flesh is taken up by seeds. In native pawpaws, the volume of seeds can exceed 50%. This variety has prolific production. Fruit tends to be small; thinning will result in larger fruit.

**Collin's Select***:* zones 5-9, grown from native pawpaws in Georgia. Half-pound fruit with a rich flavor, and mid-season ripening. Moderately prolific. Can do well in partial shade.

**Eastern Seedling***:* growing zones 5-9, these are native pawpaws dug from patches in Kentucky and Tennessee. Can serve as a pollinator for Collin's Select. Mid-season, small to medium fruit, about 30-40% seed volume. These do well in partial shade.

**Halvin***:* growing zones 5a-9, ripens mid-season, reasonably prolific, large fruit.

**KSU Atwood***:* growing zones 5a-9, early to mid-season ripening, 8% seed, highly productive. Atwood may be the most prolific of all the pawpaw varieties.

**Maria's Joy**: growing zones 5a-9. There is not a lot of information about this variety other than that it produces large fruit, which will put it mid- to later season ripening.

**Pennsylvania Golden***:* growing zones 4-9, with early ripening. Quite productive with smaller fruit, about 8% seed. Thinning could increase fruit size.

**Shenandoah***:* growing zones 4-9, with early to mid-season ripening. Moderately productive with moderately sized fruit. The flavor is slightly milder.

**Summer's Delight***:* growing zones 5a-9, with very early ripening fruit. The tree takes a little longer to start bearing fruit, around 5 years instead of 4.

These two images show the dramatic difference between the cultivated pawpaw and the native variety. You get more flesh per fruit with the cultivated varieties!

Note about thinning: If you are selling direct to consumers, then small fruit will sell as well as large fruit. If you are processing the pawpaws for freezing or other purposes, larger fruit is more beneficial as you have more pulp per fruit.

# How to grow

## Planting

You read about people foraging through woodlands for wild pawpaws. This is part of what makes pawpaw fruit rare—trees in woody, shady areas do not produce a lot of fruit. The young tree, up to a year or two years old, does best in the shade. After that, it needs sun in order to bear fruit. This is why getting a year-old tree to plant makes it a lot easier on you because you skip the shady time.

A pawpaw tree will come in a pot. They have a long taproot, which does not like to be disturbed. Dig a hole deeper than the container the tree comes in and three times as wide. You want plenty of space for the roots to spread out.

Add compost or peat to the hole when planting to improve the soil quality. Plant about 10 feet apart. Young trees can get a little sunburn on their leaves, but as the trees mature, full sun is fine. If the sun is intense in your area, provide a shade screen for the tree until it is two years old.

Pawpaw trees can grow in full shade as well as full sun. In a shady area, the tree will spread more and reach for the sky. A tree in the shade will produce fruit at the very top of the tree, making it more difficult to harvest. In a sunny area, the trees form a pyramid shape with fruit scattered throughout the tree. As always, the more sun, the more fruit.

Make sure the area drains well—standing water will kill a young or older tree.

Mulch around the tree to help protect the roots and to hold moisture in the soil.

Pawpaws need a dormant period in order to fruit.

As long as you have other flowering plants in your garden which attract pollinators, there is no problem with the trees pollinating. We do not do any hand pollinating. Ignore anyone who suggests that. However, the trees are not self-fertile. Your need two trees for best pollination, and different varieties that bloom at the same time seems to improve production. Plant the trees about 10 feet apart to ensure good pollination. If the trees are too far apart, 20 feet or more, the pollination is not as good.

Trees will start producing in 3-4 years, and produce a heavy crop after 5-6 years. The native variety can grow to 30 feet tall, and cultivars developed for fruit growing will be a bit shorter, at 15-25 feet tall.

## Fertilizer

Add compost to the soil when you plant. After that, add aged manure or compost twice a year, in early spring and then in summer. If you have soil that grows crops reasonably well, you probably do not have to do anything. We do no soil amendments for the pawpaws.

## Water

As with any young tree, water regularly during the first year to establish root growth. After that, pawpaws need water about once a month. Unless you are experiencing a prolonged drought, rain will usually be sufficient. If the leaves turn yellow early in the season, it is a sign of too much water.

### Deer resistant

Pawpaws are considered natives, which means they have survived all kinds of critters. Pawpaw trees have a slight odor which seems to be offensive to deer. At any rate, the deer seem to leave this tree alone. Now, if you have a lot of deer in your area and it is a very cold winter, they may eat it anyway. Only a fence makes an area truly deer-proof.

### Winter protection

Pawpaws do not need any winter protections. They are very hardy in growing zones 5-9, a few varieties 4-9, and are great survivors.

## Maintenance

This is a very easy to maintain fruit tree. For the first few years, there may be suckers that grow from the base of the tree. Prune them off, so the tree puts its energy into the growth of the main tree. You will get more fruit this way.

Other than that, there is very little pruning to do on the tree. It maintains its pyramid shape very well and does not produce lots of small inner branches. Prune off any broken or dead branches when you see them.

Pawpaws can handle damp soil, since they will grow near rivers, but they want good drainage. If you mound the area where you plant so that the tree sits an inch or two higher than the surrounding area, your tree will be in good shape. Pawpaws can tolerate some drought conditions, but if there is an extended period without rain, you should water. If you go four to six weeks without rain, then give the tree a drink. This is where the long taproot keeps the tree alive. It can reach any moisture which is deep in the ground.

Do not fertilize the tree for the first year. After that, you can add a top dressing of compost in the spring. Place the compost around the tree, not touching the trunk, to give the tree a boost.

The size of the fruit is dependent on how much water the tree has received, how good the soil is, how many pawpaws are on the tree, and the variety.

Do you need to thin the fruit? Thinning involves taking some of the young fruit off the tree so that the remaining pieces can grow larger. You do not have to thin if your variety achieves a moderate size. Some varieties are very prolific but with small fruit. Thinning could result in larger fruit. I choose to select varieties that have larger fruit and do no thinning at all. Most varieties do not need any thinning.

Pawpaw trees can live 30-50 years and still fruit. There are many reports of native pawpaw trees up to 75 years old and still fruiting.

# How to harvest

Pawpaws will continue to ripen off the tree, but it is best to wait until the fruit is nearly ripe. Each fruit will come off the stem easily when close to ripe. Although pawpaw skins are more durable than figs or currants, it is always best to harvest into a container that is kind to the fruit. You do not want the fruit to be piled too high, as it will bruise the fruit at the bottom of the pile. A good rule of thumb is no more than two pawpaws deep.

Our fruit is mature at the end of August and early September. The longer the fruit is on the tree, the more flavor it will have and the larger it will grow. The fruit grows in clusters of 2-9 fruit. However, not all fruit in the cluster is ready at the same time. You will harvest the pawpaws on each tree over a two-week period. Having different varieties will extend your harvest period.

You want the pawpaw to be what is called "tree ripe." This means it is fully mature, and about 1-2 days from being so soft that you cannot handle it. The pawpaw will have a yellowish-greenish cast to the skin and be slightly soft to the touch. There are small black specks on the skin.

Did you know? Pawpaws are considered berries. They snap off the stems easily without needing snips.

The unripe pawpaw is completely green, without any of the black freckles or streaks. As it ripens, it develops these darker spots on the skin. As the pawpaw ripens, it can bruise very easily.

When the pawpaw has a slight give to the touch, you can harvest it into a container. Keep the fruit in double layers only. Too much weight can cause the fruit to bruise.

If the fruit is very soft, it will bruise easily. This makes handling it difficult, even if you harvest in single layers. The skin shows every mark as it is handled, so the pawpaw starts looking like a castoff to a retail customer. A chef will recognize that the skin is discarded and the inside is what is important.

These two images show a ripe pawpaw, with some of the dark marks on the skin. The second image is a very ripe pawpaw, which will be perfect for eating immediately. Very few customers will be willing to buy it at this stage.

# Storage

For best storage and for retail sales, you want the fruit to be only slightly soft. If the fruit is just a bit soft, it will store easily for up to 2 weeks. The longer the fruit is in the refrigerator, the more likely the skin will discolor and turn black. (The fruit will be fine, but the skin will be ugly.) Once it is soft, the pawpaw should be used within a few days. The fruit will continue to ripen slowly even while refrigerated, so you want to monitor that.

If you scoop out the flesh, discard the seeds, and you can keep the pulp refrigerated for a few days. It will start to discolor after 2-3 days, so you want to use it quickly. However, the pulp freezes easily. If you want to store the pulp for any length of time, then freeze it.

Your goal is to sell the pawpaws just as they are, a little bit soft. Left out of refrigeration, they will be ready to eat in 1-2 days. If you sell at a Saturday market, for instance, leave them in refrigeration until the night before. Take them to the market and the pawpaws will be the correct softness for the customers to use that day or the next.

# How to sell

So far, demand has exceeded our supply of pawpaws. And this is without any fanfare or marketing, just putting them out in our farm market. Since the fruit has not shown up in the wholesale market reports, it is up to you to determine a price. The good news is that pawpaws seem to sell themselves. Since you will have just a few pawpaws the first year, and then 20 -30 the second year, you have the opportunity to build up demand while your tree builds up production.

How you sell is going to depend on your sales model. Do you sell to individuals, to restaurants and chefs, to wineries, to specialty stores, or to a wholesaler? Do you have a farm stand, go to farmers' markets, or set up in a business food court?

If you are selling to individuals, start with two trees. If people in your area have not tried pawpaws before, then you can sample some. However, because of the increased media presence that pawpaws already have, we have never done that. There is a pawpaw festival in our area, which does the promoting for us.

If you sell at farmers' markets, pawpaws can act as a marketing ploy as well as something to sell. It's a place to start a conversation—what are they, what do they taste like, what do you do with them. Many people have heard of pawpaws but have not eaten them before. Some people will be confused about papayas and pawpaws. Knowing the difference helps establish you as an authority. In addition to buying one or more pawpaws, customers most likely will buy something else. Plus, the people who know what pawpaws are will be thrilled. Their excitement will prompt others to try this unusual fruit as well.

If you sell to restaurants and chefs, you can approach them first before you plant. Find out if this is something they are interested in. However, since it takes several years before you get a crop, you can plant them and talk to the chefs once you have them in hand. See if they will do any cooking with pawpaws, or create a dessert with them; they may be interested in any very ripe pawpaws. In spite of your best efforts, it is all too easy to end up with some fruit that is too soft for sale at a farmer's market but is still ideal for cooking.

If you have two different chefs who are interested, others will be interested as well. You know that you will be able to sell easily in your area. Chefs at high end restaurants will definitely be interested in pawpaws. It helps the chef to let them know when you expect to have them. Good chefs are planning their menus 3-6 months in advance. Restaurants that specialize in local and fresh produce will often have daily or weekly specials into which they can incorporate unusual fruit at short notice. Any chef that is looking for ingredients that will set their restaurant apart from the others will be thrilled to have pawpaws.

It is possible that someone making wine, "shrubs," or cordials could use pawpaws, but a more likely option is pawpaw sorbet, gelato, or ice cream.

Freezing does quite well for pawpaws. Remove the seeds, and scoop the flesh from the skin. Put in containers. No sugar is needed, as pawpaws are naturally sweet. The result is similar in texture to frozen yogurt. These would sell well as a dairy-free, no added sugar, dessert. Be sure to charge for your time, the container, and the cost of holding in a freezer.

The benefit of a frozen product is the long shelf life. This also gives you a way to sell pawpaws when they are not in season. Check with your state or county regarding regulations for cottage industries for farms. Some Departments of Agriculture will serve as the inspectors and regulators.

Pawpaws do not hold up well under heat, losing much of their flavor and nuance. A pawpaw bread will be quite mild in flavor, but you could make a pawpaw-blueberry bread. Pawpaw jam does best when combined with a stronger flavored berry such as blueberry or raspberry.

If you have a dehydrator, then dried pawpaw is another way to preserve your harvest to sell at a later time. The natural sweetness is intensified in the dried state, so it is not necessary to add any sugar.

If you sell to specialty markets, then you are looking at wholesale prices. Pawpaws never show up on the USDA wholesale pricelists You will determine the price. Consider the going rate for mangoes and price accordingly. You will need to have your fruit in an appropriate sales container, such as a tray or pint or quart boxes, depending on the size of the pawpaws. Be sure to include the cost of the containers in your pricing.

Just a comment here: when you sell to chefs and restaurants, don't feel you have to discount the price unless they are getting a very large quantity. You are most likely delivering, plus providing an unusual product that they can't get elsewhere.

Be sure to check in your area for a comparable price. Reach out to a farmer who does sell pawpaws and find out what they are charging. Do not be afraid to charge a premium price. Your product may not be traveling across the country, but you are offering fresh, high-quality fruit that deserves a good price. Large pawpaws often sell for more than $1 each.

The market report I use (Philadelphia):
https://www.ams.usda.gov/mnreports/na_fv020.txt
Where you can find your local report:
https://mymarketnews.ams.usda.gov/filerepo/reports

According to USDA reports, the demand for pawpaws has been growing since 2010 and is continuing to grow. We see that in our own market. Because of that, we have never had to try to create value-added products from pawpaws.

What does that mean for you? It means you get to set the price. In general, I price the pawpaws similarly to mangoes (and no, we do not grow mangoes, but that seems to be a comparable fruit). Pawpaws have a tropical appearance, and this seems to work well for pricing. Since the price of mangoes has gone up dramatically in the past year, the price of pawpaws will also go up.

Once you get the word out, I am confident that you will sell every fruit you want.

Since you will harvest only a few pawpaws the first couple of years that the tree bears fruit, this gives you time to build an appreciative audience for them. By the time the tree has an abundant crop, you will have people waiting for them.

Note: Pawpaw skins are not edible and will cause an upset stomach if eaten.

# Propagating

Collect seeds from your fruit. Wash and clean them. Pawpaws need 90-120 days below 35 F in order to germinate. You have two options. You can plant the seeds into soil in a pot or in the ground if the temperatures are below 35 F. They have long taproots which do not like to be disturbed, so long narrow pots are preferred.

The second option is to store the seeds for 90-120 days in peat moss, and then chill in a Ziploc bag in the refrigerator. You can then plant in the spring in pots. Keep them cold (33-35 F), but do not freeze. You can sell the seeds if desired.

Once the tree sprouts, you want to keep the young tree in the shade for the first year to two. If you have mild summers, then one year is sufficient. If you have hot summers, then two years of shade is preferable.

Growing from seed has about a 10%-30% success ratio. We have chosen not to plant seeds nor to sell the young trees. It is more work than we care to do and to maintain. If we had a lot of stock that we were maintaining over the winter, it would fit in with that program. But we do not. It fits our model to order new trees as needed. That way, we can add to the number of varieties which we grow.

For the cultivated varieties, the only reliable way to propagate is by grafting.

# Persimmons

Persimmons are fabulous. They have a reputation as the "puckery" fruit, but that's only true for unripe fruit. Indeed, persimmons do have a high level of astringency which decreases as the fruit gets closer to maturity. If you eat a persimmon in its unripe state, the astringency will cause an unpleasant reaction that feels like your lips and tongue are tingling.

For decades, people thought that the Latin name, *Diospyros kaki*, meant divine fruit… but it does not. However, that does not change the fact that persimmons taste divine!

There are two types of persimmons, astringent, and non-astringent. There are three main varieties: Fuyu, Hachiya, and the American persimmon. The Fuyu type is non-astringent and the Hachiya type is astringent. They are both Asian persimmons and do well in growing zones 7-10. The American persimmon, an astringent type, grows in zones 4-10. As a native of North America, it has great hardiness.

The native American variety (*Diospyros virginiana*), which was known to indigenous tribes, was grown throughout many regions of the country. This is an astringent variety, like the Hachiyas, and contains a large seed. These persimmons were used dried or as an ingredient in bread and called "pessamin" or "putchamin" in Algonquin. The word persimmon is derived from pessamin. The seeds were roasted and used as a coffee substitute during the Civil War--even though the seeds are caffeine free.

The persimmon native to North America (*Diospyros virginiana*) is very cold tolerant, so good for very cold climates, and is also very astringent. It is suitable for growing in zones 4-10. The fruit is generally smaller (about 2-2.5 inches in diameter) than the Asian persimmons. A light frost improves the flavor dramatically and causes the fruit to sweeten.

For persimmon aficionados, the Hachiya type (*Diospyrus kaki*), shaped like an acorn, is the preferred variety. However, this type is extremely astringent with a high tannin content. Unless the fruit is completely ripe and soft, you will end up with a puckery reaction (which is fairly unpleasant) around your mouth. When you are selling to the public, this is challenging. It is too much to hope for that all of your customers are Hachiya fans. When fully ripe, the soft Hachiya is very sweet, juicy, and delicious. The astringent content has completely gone when the persimmon is completely ripe.

The other type of persimmon is the Fuyu type (*Diospyrus kaki Fuyu*). You may have seen these in the winter, marketed as Sharon fruit. These persimmons are round and flatter in shape, looking more like an orange tomato. Although the Fuyu-type persimmons do not have quite as strong a flavor as the Hachiyas, they are quite sweet. They have the distinct advantage of being far less astringent. Since the Fuyus start with a much lower astringency level, they can be eaten while still somewhat firm and at all stages up until very soft without causing a puckery reaction. For that reason, Fuyu persimmons are excellent for people who are unfamiliar with this fruit.

All persimmons are sweetest when very soft, the skin is fairly translucent and wrinkled, and you scoop the flesh off the skin with a spoon. Retail customers will not buy them at this stage. They look one step away from being spoiled. In addition, you cannot depend on your retail customers to wait until the persimmons are fully ripe before they eat them. Our solution is to have primarily Fuyu-type persimmons. We grow the Hachiya type, and we sell it clearly labeled with "Wait till completely soft" signs. Both Fuyu and Hachiya types grow well in zones 7-10.

You can find persimmons of all varieties still on the tree as leaves are dropping.

We harvest 300-500 persimmons from each tree, both Fuyu and Hachiya varieties, when the tree is fully mature. The American persimmon is as prolific, and many reports indicate even more production.

# Varieties

**Hachiya**, *Diospyrus kaki,* is the gold standard for flavor and texture among the persimmon varieties. Sweet with a soft texture, the flavor is complex and intense, and holds up well in making jam or jelly. The trees are prolific with medium to large sized fruit, about 3 inches in diameter. The fruit has a somewhat pointed end, giving the suggestion of an acorn in shape.

**Fuyu**, *Diospyros kaki Fuyu,* is the non-astringent variety of persimmons. The trees are very prolific with medium to large sized fruit and are usually seedless. The persimmons can be eaten from slightly soft to fully soft, with increasing sweetness as the fruit softens. It is a flatter shape than the Hachiya, looking more like a tomato.

**Izu** is a Fuyu derivative that ripens nearly a month earlier than its parent Fuyu. The persimmons are smaller, but the tree is quite prolific.

**American persimmon**, *Diospyrus virginiana*: this is the hardiest of the varieties for growing in North America, suitable for zones 4-9. The flavor is sweet when fully soft, the fruit is on the small side. It has an oblong shape, without the acorn point of the Hachiya.

There are now a number of American crosses with Hachiya, *D. kaki*, now called *D. virginiana x kaki*. These have the hardiness of *D. virginiana* with the flavor and larger size of the Hachiya, *D. kaki*.

Some of the desirable hybrids include Claypool, Yates, Celebrity, and Lehman's Delight. These crosses have fewer seeds than the American persimmons.

There are also some very late producing trees of the American-Hachiya crosses with names such as Deer Magnet and Deer Candy. These ripen in December, after the leaves have fallen, and are a favorite for deer to come visit and eat. The fruit is still sweet for human consumption, but smaller and matures very late in the season.

All of the varieties are extremely soft when fully ripe, like a jelly with a skin around it. The skin will look wrinkled and almost translucent, sometimes with dark spots on the skin. Retail customers are very cautious and reluctant to purchase when the persimmons reach this stage.

# How to grow

## Planting

Allow plenty of space between plants as the trees can grow to substantial height and width, about 20-25 feet apart for Fuyu or Hachiya types. These trees need full sun, farmer's definition (sun up to sundown). Less than full sun results in leaves and no fruit.

The Fuyu persimmon will grow to about 15 feet tall and a 15-foot canopy;

The Hachiya can grow to about 20-30 feet tall with a 30-foot canopy;

The American persimmon and American crosses can grow 30-50 feet tall, with a 50-foot canopy.

If you have a small space, look for smaller varieties, such as the Fuyu and Fuyu crosses. Izu is a Fuyu cross that is only about 10-12 feet tall. The other option is to prune the top growth to force the tree to grow laterally. This, however, defeats the goal of low maintenance.

The tree has a long taproot to help stabilize the significant top growth, so allow room for the tree to grow in all directions. Remember that the roots take up as much space in the ground as the canopy of branches and leaves above the ground.

Dig a hole twice as wide as the root ball and slightly deeper than the tap root. Soak the plant in a bucket of water for two hours before planting. Fill in the soil around the plant. Spread a layer of compost around the base of the tree, but not touching the trunk.

Use two stakes to stabilize the young tree, anchoring it with tree ties, or a hose slipped over twine or wire. Whatever touches the tree trunk cannot gouge into the bark or damage it. After 2-3 years, the tree will not need the stakes, and they can be removed.

Persimmon trees are self-fertile, but it is still recommended to have two or more trees for better pollination (people will love them so much, you will want at least two!). We had one tree for several years and sold out very quickly. We have planted 10 more.

### Water

Water the young trees once a month deeply. Remember that the tap root is reaching two feet or more into the ground, so you want the water to get to the full root system. Trees take in water through their roots and spread it through the trunk and branches from there.

After the first year, you only need to water if there is a prolonged heat spell or drought in your area. Rain once a month will be sufficient. If it does not rain, then you need to water.

**Fertilizer**

Add compost as a top dressing when you plant, and then once each year in the early spring. If you do not have compost, use a balanced fertilizer, such as 10-10-10, with equal proportions of magnesium, nitrogen, and potassium. Water and adequate nutrients help determine the size of the fruit, so feeding the tree each year helps with your yield. If you forget, you will still have a good crop.

**Pruning**

Persimmons do not need a lot of work. Trim any dead branches off. Each spring, you can trim out any small branches or any that are crowding or rubbing into another branch. The tree produces fruit on long branches, so any short branches can be pruned off. We find we need to do very little pruning.

**Life span**

How long will persimmon trees live? The oldest documented persimmon tree is a Hachiya in Japan, at 650 years old. Clearly that tree is happy in its location. The astringent varieties live longer than the non-astringent.

Persimmon trees will start bearing fruit after 3-4 years, but are about 6 years old before they have a full crop. A mature tree of about 10 years old can produce about 300 pounds of fruit each year. The Fuyu tree will produce heavily for about 10 years, with a life span of 15 to 20 years. The Hachiya tree will live for 40-60 years, producing heavily after it is 10 years old. The American persimmon tree can live to be over 100 years old and produce up to 500 pounds of fruit per tree.

You can see that the astringent varieties are longer-lived than the non-astringent. We grow some of each, recognizing that we have to replace the Fuyu-type trees more frequently. We want the non-astringent variety available for our customers who are not as familiar with persimmons. It makes it easy for them to try and enjoy this fruit.

Because the Hachiya or American tree can grow so tall (30 to 50 feet or more), ladders are needed for harvesting. You can also top cut the tree to limit its height, and it will produce more lateral growth. If you want to keep the tree shorter, you need to start that pruning from the beginning.

There are pros and cons to each variety, so you need to select which will work best for you.

# Harvest

When the fruit is orange, you can harvest. Persimmons will continue to ripen off the tree. They are fully ripe when the skin turns darker, with a translucent quality, and the fruit is very soft. You do not want to wait to harvest until the fruit is soft, as it can bruise very easily and is much more difficult to handle. The fruit will be orange on the tree for several weeks, so you want to leave it as long as possible to achieve maximum size (maximum price!). This also gives you a long harvest window, so you do not have to harvest on a particular day. Persimmons rarely have a problem with splitting due to rain.

The Fuyu persimmons have the advantage of a range of textures for eating. When they are firm, they can be similar to an apple, or a little soft and they are similar to a peach texture. When very ripe, they are similar to jelly in texture, as are the astringent Hachiyas and Americans. The astringent varieties can only be eaten when soft, but you can harvest them before they are fully softened. Oftentimes, it is the texture of the fruit that will determine how much people will like it.

If you leave persimmons to ripen on the tree, wild turkeys, raccoons, squirrels, and deer will visit. All the more reason to harvest while firm.

Harvest by snipping the stem right above the calyx (the leaves sitting on the crown of the fruit). You want to leave the calyx intact. If you try to take it off, it opens the top of the persimmon and exposes the soft flesh. It will start to decay quickly. You want to snip as close to the calyx as possible so that any stem will not poke into another persimmon and puncture the skin.

The calyx stays the same size and the persimmon grows. By maturity, the leaves are quite small proportionately. Leave them attached to the persimmon when harvesting to help with storage.

In the firm stage, you can fill a harvest basket without damaging any of the fruit. Handle gently, because they will show bruising when they soften.

Persimmons are typically harvested in October-November, making them a very late fruit.

For taller trees, you will need a ladder to capture the fruit at higher levels. Eventually, the fruit will drop, when it is completely ripe. It is extremely soft at that point and not for sale. Fruit that has dropped on the ground must be discarded according to GAP, Good Agricultural Practices. You may opt to harvest the lower levels and leave the higher fruit. When it drops, you can plant the seeds to grow more trees. You can salvage fruit for your own use, such as for animal feed, but drops cannot be sold.

Larger trees have a longer lifespan, so you can weigh the benefits of a longer life and harder to harvest vs. a shorter tree and easier to harvest. If you are in a colder zone, you may decide that the hardier variety will be better for you. The volume of fruit produced makes up for the difficulty of harvesting the upper part of the tree. If you don't get it all, it's fine! Your other option is to prune the central trunk each year to make the tree branch out and grow laterally. You can have the tree grow along a fence or have some supports if you do that. Because we have an extra-long list of spring work, we choose to have slightly smaller trees and ignore pruning.

To grow Fuyu or Hachiya persimmons in zone 6, find an area protected from winter winds and an area that frosts later than other parts of the property. The southern part of Zone 6 will be more successful than the northern part. A smaller tree can be given winter protection with row cover, an unheated tunnel, or wrapping in burlap. The American persimmon is hardy to zone 4, so if you do not want to bother with winter protection, grow this variety. One of the American-Hachiya hybrids will bear more reliably in zones 4, 5, and 6 than a non-hybrid Hachiya or Fuyu.

Again, pros and cons. You need to choose based on what works for your location, the space that you have, and the time available that you have at each season.

# Storage

In the firm stage, in refrigeration, persimmons will keep for 4-6 weeks. The persimmons will continue to slowly ripen even in refrigeration. Once the persimmons are at room temperature, they will soften within a few days. When fully soft and ripe, persimmons only keep 3-4 days, even if refrigerated at that stage.

Store at 33-36 F, away from apples. Apples naturally produce ethylene, which hastens the ripening process. Other than that, just keep them refrigerated. No special containers are needed.

The long storage life gives you time to sell all of your crop without feeling pressured.

As the persimmons soften, the calyx, the top green leaves attached to the persimmon, will loosen. When the persimmon is fully ripe and soft, the calyx will come off with very little pressure. If a calyx has come off the persimmon while in storage, you should freeze, dehydrate, or turn it into jam as soon as possible. Or eat it yourself! You should eat your fair share of these wonderful fruits.

If you are dehydrating any of the persimmons, cull out any that are too soft for sale but still good, and put those in the dehydrator. They will have maximum flavor and sweetness at that point.

# How to sell

Persimmons are highly prized and easy to sell. Once people experience persimmons, they are ready for more.

Farmers' markets, market hubs, and on-farm stands make persimmons easy to sell. Chefs can also be a good way to sell persimmons.

If people are not familiar with persimmons in your area, then try samples. Give one to people who purchase a certain amount, or add to your CSA shares.

To freeze, scoop out the ripe persimmon flesh from the skin, and freeze in the size container you will want to use (cup, half-cup, etc.). Eating it frozen is like having sorbet! You can sell frozen persimmon as a dessert, maintaining the proper temperatures. Check your local regulations regarding cottage industries.

Persimmons also make excellent jam (now a customer favorite here). We don't always have enough persimmons unsold to turn into jam, but when we do, people line up to buy the jam. It makes a beautiful, rich, orange-colored jam.

To price your persimmons, you can check the wholesale market reports. Sometimes persimmons will show up. You should, at the very least, double the wholesale cost. We price our larger persimmons similarly to pawpaws or slightly less than mangoes. The smaller persimmons are priced similarly to large figs. We nearly always sell by the piece. If there is a quantity of smaller persimmons, we can put 5 or 6 in a quart box and sell that way.

If you have seedlings in the spring, this is another way to sell your persimmons. Note: Fuyu persimmons are mostly seedless, so grafting is the only reliable way to start some new plants.

The market report I use (Philadelphia):

https://www.ams.usda.gov/mnreports/na_fv020.txt

Where you can find your local report:

https://mymarketnews.ams.usda.gov/filerepo/reports

# Propagating

You can save the seeds from an existing fruit and plant that to grow some more trees. The Hachiyas and American persimmons have seeds that you can save.

The Fuyu persimmons are mostly seedless, so grafting is the preferred method for propagating the Fuyus.

The American-Hachiya hybrids need to be grafted as the seeds will not grow true to type.

To grow from seed, you will need to mimic the natural dormant stage of the plant. Save the seeds from the completely ripe fruit, clean all flesh off, then soak the seeds for three days in warm water. Rinse off to remove any remaining fresh. Wrap in a damp paper towel and place in a glass container. Any clean glass jar will work. Leave it uncovered; the paper towel needs to be accessible. Place in the refrigerator. Leave the jar in your refrigerator for about three months. Keep the paper towel moist by spritzing it with water when it dries out.

After three months, plant your seeds in tall pots so that the taproot can develop. You want a warm location of about 70 F. Plant the seeds about two inches deep in the soil, with about eight inches of soil below the seed. It's a good idea to plant several seeds as the germination rate for persimmon seeds is about 15%-25% success. You will see sprouts in 6-8 weeks.

Keep the soil moist but not soggy wet. Keep the plant in bright sunlight. When the last frost is past, put the potted plant in a semi-shady area to help it acclimate to being outside. After the two weeks, move the plant(s) into a sunny area. The best time to transplant into the ground will be in the fall, in October or November.

If you can transplant them into the ground after a good, soaking rain, it will be easier for the roots to establish in the ground. Be sure to mulch to protect the roots.

If starting more seedlings fits with the model of what you are doing on your farm, then this can be an exciting addition to a plant propagation program. Our choice is to order additional trees, as our to-do lists are longer than our time available.

# Calendar

What does the year look like for most growing zones with these five fruits? It's not complicated.

**January**: Rest.
**February**: Rest.
**March**: Prune out dead canes in currants and gooseberries. Divide any gooseberries or currants if desired.
**April**: Do the pruning for the gooseberries and currants if you didn't get it done in March. Add a top dressing of compost to all of the plants (currants, gooseberries, figs, pawpaws, and persimmons). Add a fresh layer of mulch. Dig any fig roots for division if desired.
**May**: Uncover the fig trees. Prune out any dead branches. Add a fresh layer of mulch.
**June**: Harvest and sell currants and gooseberries. Harvest and sell light "breba" crop of figs, if any.
**July**: Harvest and sell late varieties of currants and gooseberries. Add some compost around your fig trees.
**August**: Pay attention to the rest of your farm until the end of August, when figs and pawpaws may start.
**September**: Harvest and sell figs and pawpaws.
**October**: Harvest and sell figs and persimmons.
**November**: Once the leaves have dropped, prune your figs to a desired height and cover them for winter protection, if you are in a colder zone. Take cuttings of currants or gooseberries if desired. Set aside your profits.
**December**: Rest.

# A few notes about value-added products

In general, creating value-added products requires time, hence the name. Value-added products usually require their own specific supplies, systems, space, and knowledge. For instance, if you are making jam, it requires specific cooking implements, thermometers, jars, lids, a licensed kitchen, recipes, and labels. Essentially you are setting up another business. That means tracking the time involved as well as the specific costs required for the project.

It's why we do as little value added as possible. First, we prefer to sell the fruit fresh in our existing sales venues. Second, we do not want to put the time into creating a value-added product. We partner with a small business whose specialty is making jam. The owner has the equipment, licensed kitchen, and skills to make jam efficiently. We can send persimmons to her and she sends back persimmon jam. We have a unique product without having to do the extra work.

Propagating fruit trees from seed requires chilling space and attention to the seeds during that time. Nurturing young seedlings requires nearly daily attention as well as specific lighting and heat requirements. It is a routine that does not fit with what we currently do. While "tipping" is relatively easy, when you have multiple employees, it is easy for someone to forget what is happening and accidentally cut down or otherwise negate that work.

Dividing plants at the root is much quicker and easier. The young plants are very hardy and transplant into pots or other areas without any fuss. This is a process we do when there is time. The beauty of this is that if we are unable to do it one spring, we can always do it the next year.

This is not to discourage you from creating value-added products. Value-added products may fit your business model quite well. Do consider working with a strategic partner who can do the work for you and give you a finished product that is ready to sell.

Look at where you want to spend your time. For us, growing is what we like best and we prefer to focus as much of our time on growing as we can. For other people, creating jams or cordials or dried fruit is the fun, intriguing activity. Work with your interests and strengths.

# Conclusion

Can you make money growing on a small farm? Yes, you can.

I speak from my experience as a farmer who grows fruits and vegetables on a small acreage farm—just 5.5 acres—and we generate enough revenue to support three generations. About three acres are in production. We sell directly to the customer; we do not wholesale.

The better question to ask is, can you make a profit? It's not how much money you make; it's how much you are able to keep.

How can you do this?

First, you have to make the transition from a home gardener to a farmer. The home gardener grows for the home and for enjoyment. The farmer is running a business and has the goal of bringing money into the business. This can be a modest goal as a part-time farmer, such as paying for vacations or paying off the mortgage. It does not have to be a full-time profession, although it can be. Start small and increase your planting and selling. See how well you like all aspects of planning, planting, harvesting, and sales.

Second, you need ground that is capable of growing crops. Have you grown anything in this space before? What kind of soil is it? If you have never grown anything there before, it may take a couple of seasons to get compost in the ground and make it a growing area.

Third, you need to know the rhythm of the seasons in your area. What are your frost-free dates, spring and fall? How much rain do you get? Is there a dry spell in the summer? What is your growing zone? This affects what you can grow easily in your area.

## Here are five points to consider

### Grow what you are able to sell.

Find the best market. If it's a farmers' market, you want lots of people who are ready to buy from small farmers. I have gone to farmer's markets where only a handful of people come by, they aren't ready to buy, or they are looking for a half-priced bargain. That's not what you want! Other markets have lots of customers who are ready to buy and want to support small farms.

If it's a stand on your property, you want lots of traffic so that enough people see your stand and will buy what you have. If it's through an online food hub, you want one with a substantial number of existing customers who are ready to buy what you have. If you sell to chefs, you want restaurants that will buy in the volume that you have to sell.

There is no point in growing figs for sale if everyone has a fig tree in their backyard. There is no point in growing ten rows of currants if the potential customers in your area will only buy all the currants from one row. Every market is different. You need to know what your market wants.

**Grow what will bring you the most amount per square foot per growing season.**

Why settle for $2 per foot if you can get $8 per foot by growing a different crop? Remember #1, you have to be able to sell it.

**What can you grow that others don't grow at all?**

These fruits are suggestions of unusual crops that most farmers miss. This is a way to have a unique selling proposition.

**Can you grow or produce anything that keeps well?**

This way you have something to sell even if the fresh crops aren't ready yet or you have sold out of lettuce and tomatoes. Persimmon jam, fig preserves, dried pawpaws, currant syrup, dried currants, gooseberry jam? Value-added products are called that for many reasons. In addition, these crops will keep well in refrigeration, giving you time to sell them as a fresh crop.

**Have more than one sales venue to sell what you grow.**

Sell at a farmers' market, sell through a Community Supported Agriculture program, sell to restaurants, sell at a home market, and sell through an online ordering system. It is good to have at least two methods of selling your crops.

Even after farming for decades, these are the rules that I use to evaluate any new crop. I also use these rules to evaluate what the farm currently grows, to see if we can improve our efficiency, productivity, and profitability. As the market changes, we have to adapt. The basic rules for evaluating crops remain the same: Can we sell it? At what price? What are the costs in time and supplies? Does it add anything extra to what we grow?

You can grow a lot on a few acres of ground and bring in a good income. The trick is to plan it out for maximum revenue while controlling time and expenses. This means you have to look at the sales end first. Where and how you sell your produce makes a difference. If you have to make multiple deliveries, that cuts into your working time and your profitability.

You want a venue that fits your products and has the volume of people you need to sell all of your products. If you have 50 pints of figs each week, it does not work if you can only sell 25 of them at the farmers' market. Can you sell the rest of the figs through online orders? Study the different markets and see how much the farmers actually sell.

If someone has a beautiful display of homemade preserves but only sells 20 jars, they didn't make enough to pay for their time, the market fee, and the cost of goods.

Look for the market where the farmers bring a full van or truckload of products and leave with nothing. That's the market that you want.

Can you grow lots of produce on a few acres? Yes, you can. The trick is selling the produce and keeping a good profit for yourself.

This is where these five fruits help your profitability. They take up very little time during the year, except to harvest and to sell. They often can be planted where other crops do not grow.

Think about what suits your growing climate and your customers.

Which of these fruits fit in with what you are currently doing? How do they complement your current crops and what your current customers like to get from you?

Do people come to you for fresh-cut herbs? Fresh herbs and leaf lettuces are in the unusual greens categories at markets. Unusual fruits will fit right in.

Do you sell your own cheese? Fruit and cheese are a classic combination.

Do you sell plants? Then propagating and selling your own fruit shrubs and trees could be a natural expansion. Keep in mind that trees and shrubs take up more space, need to be in larger pots, and generally take longer to sell. That means you will have more weeks of maintaining them in the pots before they are sold.

Do you sell jarred goods—jams, pickles, salsas? Any of these fruits are a great addition and provide unusual flavors that can set your line apart from others. Having fresh fruit and then later having the jams make for a great combination.

Look for crops that fit in with what you are doing, with the equipment that you have, and with the schedule that you want to maintain. The goal is to improve your profit without increasing your work schedule. No one needs to work more hours. We want these trees and shrubs to work for us. Fortunately, they do not need specialized equipment. A shovel to dig the hole to plant and water for the young plants take care of most everything.

In general, you want to work with the growing seasons and not against them. We are in zone 7a, and it would be a lot of work for us to harvest figs routinely in early May. Is it possible? Yes. Would it be profitable? Unlikely. The heating and lighting requirements would cost more than what we could get in sales.

We know that we can grow gooseberries, currants, figs, persimmons, and pawpaws with very little effort. Figs are the only crop that we have to do a little bit of work on in the fall. Make the weather work in your favor. Choose crops that flow with your weather.

I hope this gives you a new perspective on how to evaluate crops that you are growing to bring income to your home, homestead, or farm. Whether you are looking at a side hustle or a full-time living from farming, then fruit can help you achieve your goals.

By no means is this an exhaustive list of low-maintenance fruit. These are examples and are designed to make you think about what will fit in with what you are doing. However, these are fruits that reliably bear a crop year after year. They are not bothered by late frosts, dry spells, or rainy weather. They are resistant to diseases, they don't get mold, and insects don't bother them. We love these crops for many reasons!

**When you are looking at something to grow and you want to make some money from it, consider these three points.**

- Does this fit in with what you already do? Do you have the equipment on hand to take care of this type of crop? Do you have the space for it?
- How much time does it require to maintain this crop? Your time is valuable. If you have to provide maintenance all year before you get to harvest and then sell, then you have a lot of time invested in that crop. These are crops that you can pretty much ignore until you harvest and sell, which preserves your time for other projects.
- How many years will you be able to harvest? What's the return on your investment of time and space? Many years, indeed, for these fruits. If you are looking to move in a few years, then planting trees and shrubs may not be in your best interest right now.

Fruit can be the marketing tool that brings customers to you. At farmers' markets, we had people rushing at the beginning of the market so that they could get figs or persimmons or gooseberries. While they were there, then they also picked up their lettuce and carrots and scallions. It's great to have marketing pay for itself plus bring in some extra profit.

These are also crops that are not seen in grocery stores, so they emphasize the fact that you are a grower. Few farms are growing these crops, so it helps distinguish you from anyone else in the area who is only growing vegetables.

If you have an email list, these are also great crops to let your customers know are now available and then you can have a link for an easy recipe in your email.

I am committed to helping farmers make a profit. We feed our communities, and our communities are better for having us in their midst. But we need money to pay our bills, and selling what we grow is how we do it. I hope this helps you with bringing profit home and supporting your growing farm business.

# Author's note

Dear Reader,

Thank you for your purchase of this book!

This book has grown out of our experiences growing and selling these crops on our small family farm. We love how dependable these fruits are, that we have a crop no matter what the weather is doing. We love how easily we are able to sell these fruits. I hope that all of these crops prove as profitable for you as they continue to be for us.

If this book has been helpful, please take a minute and leave a review on Amazon. This helps the book be shown to more people and helps support our farm. My goal is for the farm to be here for another 190 years.

Happy growing and happy selling!

~ Ruth

# About the author

Ruth Linton grew up and worked on her family farm, then operated her own business for ten years before returning to the home farm. She is the sixth generation of her family to run Highland Orchards in Wilmington, Delaware, a small acreage farm that is busy with a year-round on-farm market, a year-round Community Supported Agriculture program, an online ordering part of the business, and local restaurant sales. She took produce to nearby farmers' markets for twenty years.

Prior to Ruth's ventures in going to farmers' markets from 1999-2019, generations of family members went to the King Street Farmers' Market in Wilmington from 1832-1955.

Ruth has numerous online courses available. When she isn't writing or creating a new course, she is talking with her customers at the farm.

Sign up for a weekly email to keep up with what's new in books and courses, and what's happening on the farm. Follow Highland Orchards on Facebook or Instagram to see what's growing.

Made in the USA
Monee, IL
05 November 2026

67378081R00057